Lesson Activity Book

Developed by Education Development Center, Inc.
through National Science Foundation
Grant No. ESI-0099093

Published and distributed by:

www.Math.SchoolSpecialty.com

Copyright © 2008 by Education Development Center, Inc.

All rights reserved. No part of this publication may be reproduced or transmitted in any form or by any means, electronic or mechanical, including photocopy, recording, or any information storage and retrieval system, without permission in writing from the publisher.

Requests for permission to make copies of any part of the work should be addressed to Permissions and Copyrights, School Specialty Math, 80 Northwest Boulevard, Nashua, NH 03063.

Think Math! Lesson Activity Book, 2
Printing 15— 5/2017
Webcrafters, Madison, WI

1358078
978-1-60902-256-3

If you have received these materials as examination copies free of charge, School Specialty Math retains title to the materials and they may not be resold. Resale of examination copies is strictly prohibited and is illegal.

Possession of this publication in print format does not entitle users to convert this publication, or any portion of it, into electronic format.

This program was funded in part through the National Science Foundation under Grant No. ESI-0099093. Any opinions, findings, and conclusions or recommendations expressed in this program are those of the authors and do not necessarily reflect the views of the National Science Foundation.

Principal Investigator
E. Paul Goldenberg

Curriculum Design and Pedagogy Oversight

E. Paul Goldenberg	Lynn Goldsmith	Nina Shteingold

Research

Director: Lynn Goldsmith

Nina Arshavsky	Sabita Chopra	Suenita Lawrence
Cynthia Char	Sophia Cohen	Katherine Schwinden
	Andrea Humez	Eugenia Steingold

Editorial

Director: Frances Fanning   Nicholas Bozard   Eric Karnowski

Writing

Director: Eric Karnowski

Jean Benson	Stacy Grossman	Paisley Rossetti
Abigail Branch	Andrea Humez	Nina Shteingold
Sara Cremer	Suenita Lawrence	Kate Snow
E. Paul Goldenberg	Debora Rosenfeld	Julie Zeringue

Graphics and Design

Directors: Laura Koval and Korynn Kirchwey

Jessica Cummings	E. Charles Snow
Jennifer Putnam	Jenny Wong

Project Management

Directors: Eric Karnowski and Glenn Natali

Amy Borowko	Alexander Kirchwey	Kimberly Newson
Nannette Feurzeig	Helen Lebowitz	David O'Neil
Kim Foster	June Mark	Cynthia Plouff

Mathematics Reviewers

Richard Askey, Professor of Mathematics, Emeritus
University of Wisconsin, Madison, Wisconsin

Harvey Keynes, Professor of Mathematics
University of Minnesota, Minneapolis, Minnesota

Roger Howe, Professor of Mathematics
Yale University, New Haven, Connecticut

David Singer, Professor of Mathematics
Case Western Reserve University, Cleveland, Ohio

Sherman Stein, Professor of Mathematics, Emeritus
University of California at Davis, Davis, California

Additional Mathematics Resource

Al Cuoco, Center Director, Center for Mathematics Education, Education Development Center, Newton, Massachusetts

Advisors

Peter Braunfeld	June Mark
David Carraher	Ricardo Nemirovsky
Carole Greenes	James Newton
Claire Groden	Judith Roitman
Deborah Schifter	

Evaluators

Douglas H. Clements	Mark Jenness
Cynthia Halderson	Julie Sarama

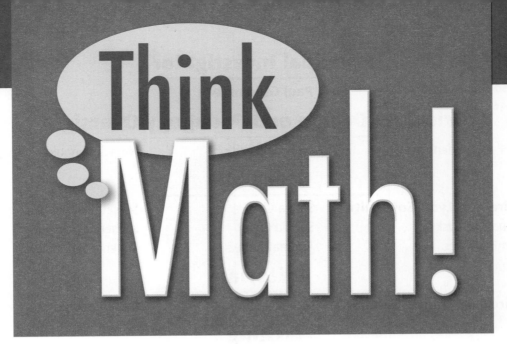

Chapter 1 Counting Strategies

Chapter Investigation 1
School-Home Connection 2
Lesson 1   Repeating and Growing Patterns 3
Lesson 2   Working with Number Patterns 5
Lesson 3   Writing Number Sentences 7
Lesson 4   Adding and Subtracting on the Number Line 9
Lesson 5   Completing Number Sentences 11
Lesson 6   Skip-Counting on the Number Line 13
Lesson 7   More Skip-Counting on the Number Line 15
Lesson 8   Systematic Counting 17
Lesson 9   Finding Ways to Make 10 19
Lesson 10 Previewing Multiplication, Part I 21
Lesson 11 Previewing Multiplication, Part II 23
Lesson 12 Problem Solving Strategy: *Look for a Pattern* 25
Problem Solving Test Prep 26
Chapter Review/Assessment 27

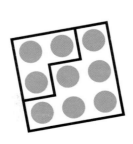 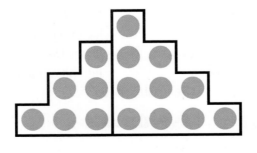 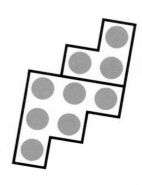

Chapter 2   Working with 10

Chapter Investigation	29
School-Home Connection	30
Lesson 1 Finding Sums of 10	31
Lesson 2 Introducing > and <	33
Lesson 3 Finding and Comparing Sums to 10	35
Lesson 4 Mastering Sums of 10	37
Lesson 5 Adding and Subtracting with 10	39
Lesson 6 Finding How Close to 10	41
Lesson 7 Adding Numbers by Making 10	43
Lesson 8 Rounding to the Nearest 10	45
Lesson 9 Problem Solving Strategy: *Solve a Simpler Problem*	47
Problem Solving Test Prep	48
Chapter Review/Assessment	49

Think Math! Contents

Chapter 3 Place Value

Chapter Investigation		51
School-Home Connection		52
Lesson 1	Estimating and Counting Larger Numbers	53
Lesson 2	Grouping by Tens and Hundreds	55
Lesson 3	Representing Two-Digit Numbers	57
Lesson 4	Representing Three-Digit Numbers	59
Lesson 5	Regrouping	61
Lesson 6	Using Place Value to Compare	63
Lesson 7	Connecting Numbers and Words	65
Lesson 8	Working with Hundreds, Tens, and Ones	67
Lesson 9	Problem Solving Strategy: *Draw a Picture*	69
Problem Solving Test Prep		70
Chapter Review/Assessment		71

Chapter 4 Addition and Subtraction with Place Value

Chapter Investigation		73
School-Home Connection		74
Lesson 1	Exploring Addition with Base-Ten Blocks	75
Lesson 2	Exploring Subtraction with Base-Ten Blocks	77
Lesson 3	Wonder Wheel Addition and Subtraction	79
Lesson 4	Introducing the Cross Number Puzzle	81
Lesson 5	Addition and Subtraction	83
Lesson 6	Adding a Multiple of 10	85
Lesson 7	Fewest Dimes and Pennies	87
Lesson 8	Fewest Dollars, Dimes, and Pennies	89
Lesson 9	Problem Solving Strategy: *Work Backward*	91
Problem Solving Test Prep		92
Chapter Review/Assessment		93

Chapter 5 Probability and Data

Chapter Investigation	95
School-Home Connection	96
Lesson 1 Exploring Probability	97
Lesson 2 Using Real-Object Graphs and Picture Graphs	99
Lesson 3 Using Bar Graphs to Investigate Probability	101
Lesson 4 Making and Using Bar Graphs	103
Lesson 5 Making and Using Pictographs	105
Lesson 6 Graphing Change Over Time	107
Lesson 7 Problem Solving Strategy: *Make a Table*	109
Problem Solving Test Prep	110
Chapter Review/Assessment	111

Chapter 6 Measuring Time

Chapter Investigation	113
School-Home Connection	114
Lesson 1 Exploring Time	115
Lesson 2 Minutes in an Hour	117
Lesson 3 Telling Time to 10 Minutes	119
Lesson 4 How Far? How Fast?	121
Lesson 5 Telling Time to 5 Minutes	123
Lesson 6 Telling Time to the Minute	125
Lesson 7 Calendar and Ordinal Numbers	127
Lesson 8 Problem Solving Strategy: *Look for a Pattern*	129
Problem Solving Test Prep	130
Chapter Review/Assessment	131

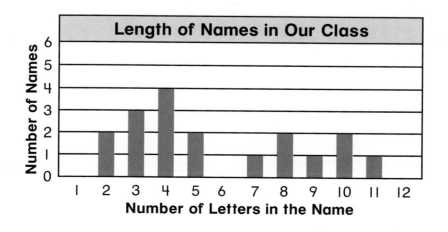

Contents

Chapter 7 Doubling, Halving, and Fractions

Chapter Investigation	133
School-Home Connection	134
Lesson 1 Exploring One Half	135
Lesson 2 Finding Half: Even or Odd	137
Lesson 3 Doubling Numbers	139
Lesson 4 Halving and Doubling Time and Numbers	141
Lesson 5 Doubling Length	143
Lesson 6 Thirds and Fourths	145
Lesson 7 Fair Shares	147
Lesson 8 Exploring Fractions with Cuisenaire® Rods	149
Lesson 9 More Fractions	151
Lesson 10 Problem Solving Strategy: *Guess and Check*	153
Problem Solving Test Prep	154
Chapter Review/Assessment	155

Chapter 8 Building Addition and Subtraction Fluency

Chapter Investigation	157
School-Home Connection	158
Lesson 1 Adding with Cuisenaire® Rods	159
Lesson 2 Exploring Fact Families	161
Lesson 3 Connecting Addition and Subtraction	163
Lesson 4 Adding and Subtracting Using 5 and 10	165
Lesson 5 Adding and Subtracting Numbers Near 10	167
Lesson 6 Place Value and Cross Number Puzzles	169
Lesson 7 Breaking Numbers Apart	171
Lesson 8 Using Cross Number Puzzles to Subtract	173
Lesson 9 Comparing Mathematical Expressions	175
Lesson 10 Creating and Solving Story Problems	177
Lesson 11 Strategies for Multiple-Choice Questions	179
Lesson 12 Problem Solving Strategy: *Solve a Simpler Problem*	181
Problem Solving Test Prep	182
Chapter Review/Assessment	183

Chapter 9 Two-Dimensional Figures and Spatial Sense

Chapter Investigation		185
School-Home Connection		186
Lesson 1	Sorting Polygons by Attributes	187
Lesson 2	Congruent and Similar Figures	189
Lesson 3	Building with Triangles	191
Lesson 4	Looking at Reflections	193
Lesson 5	Lines of Symmetry	195
Lesson 6	Cutting Polygons Apart	197
Lesson 7	Measuring Area	199
Lesson 8	Recording Paths	201
Lesson 9	Directions from Here to There	203
Lesson 10	Problem Solving Strategy: *Draw a Picture*	205
Problem Solving Test Prep		206
Chapter Review/Assessment		207

Chapter 10 Adding and Subtracting Larger Numbers

Chapter Investigation		209
School-Home Connection		210
Lesson 1	Making Sums of 100	211
Lesson 2	Adding with Coins	213
Lesson 3	Patterns in Money	215
Lesson 4	Place Value in Money	217
Lesson 5	Computing with Money	219
Lesson 6	Adding Two-Digit Numbers	221
Lesson 7	Subtracting Two-Digit Numbers	223
Lesson 8	Exploring Expanded Notation	225
Lesson 9	Mental Math with Three-Digit Numbers	227
Lesson 10	Adding Two- and Three-Digit Numbers	229
Lesson 11	Subtracting Two- and Three-Digit Numbers	231
Lesson 12	Practice Adding and Subtracting	233
Lesson 13	Problem Solving Strategy: *Solve a Simpler Problem*	235
Problem Solving Test Prep		236
Chapter Review/Assessment		237

Think Math! Contents

Chapter 11   Skip-Counting and Equivalent Sets

Chapter Investigation... **239**
School-Home Connection **240**
Lesson 1   Looking for Patterns in Jumps............................ **241**
Lesson 2   Combining Equivalent Sets................................. **243**
Lesson 3   Organizing Equivalent Sets................................. **245**
Lesson 4   Adding Equivalent Sets....................................... **247**
Lesson 5   Working with Rectangular Arrays....................... **249**
Lesson 6   Building Multiples **251**
Lesson 7   Sharing Between Two Children **253**
Lesson 8   Sharing Among Three Children **255**
Lesson 9   How Many Packages? **257**
Lesson 10 Problem Solving Strategy: *Make a List*.................. **259**
Problem Solving Test Prep **260**
Chapter Review/Assessment.. **261**

Chapter 12   Measuring Length

Chapter Investigation... **263**
School-Home Connection **264**
Lesson 1   Measuring Length with Nonstandard Units **265**
Lesson 2   Measuring to the Nearest Inch............................ **267**
Lesson 3   Measuring in Inches, Feet, and Yards **269**
Lesson 4   Relating Inches, Feet, and Yards........................ **271**
Lesson 5   Using Fractions to Measure Length..................... **273**
Lesson 6   Measuring to the Nearest Centimeter **275**
Lesson 7   Measuring in Centimeters and Meters **277**
Lesson 8   Problem Solving Strategy: *Act It Out*.................... **279**
Problem Solving Test Prep **280**
Chapter Review/Assessment.. **281**

 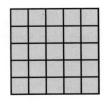 

Chapter 13 Exploring Multiplication and Division

Chapter Investigation	283
School-Home Connection	284
Lesson 1 Counting Combinations	285
Lesson 2 Counting Intersections	287
Lesson 3 Counting Hidden Intersections	289
Lesson 4 Introducing Division	291
Lesson 5 Multiplication and Division Fact Families	293
Lesson 6 Multiplication and Division Models	295
Lesson 7 Dividing and Estimating with Coins	297
Lesson 8 Problem Solving Strategy: *Guess and Check*	299
Problem Solving Test Prep	300
Chapter Review/Assessment	301

Chapter 14 Comparing and Contrasting Three-Dimensional Figures

Chapter Investigation	303
School-Home Connection	304
Lesson 1 Two- and Three-Dimensional Figures	305
Lesson 2 Faces	307
Lesson 3 Edges	309
Lesson 4 Vertices	311
Lesson 5 Cylinders and Cones	313
Lesson 6 Problem Solving Strategy: *Make a Table*	315
Problem Solving Test Prep	316
Chapter Review/Assessment	317

Contents

Chapter 15 Capacity, Weight/Mass, and Temperature

Chapter Investigation	319
School-Home Connection	320
Lesson 1 Comparing, Ordering, and Measuring Capacity	321
Lesson 2 Measuring in Cups, Pints, Quarts, and Gallons	323
Lesson 3 Measuring in Milliliters and Liters	325
Lesson 4 Comparing and Measuring Weight	327
Lesson 5 Measuring in Grams and Kilograms	329
Lesson 6 Measuring in Ounces, Pounds, and Tons	331
Lesson 7 Measuring Temperature	333
Lesson 8 Problem Solving Strategy: *Act It Out*	335
Problem Solving Test Prep	336
Chapter Review/Assessment	337

Chapter 16 Multiplying and Dividing

Chapter Investigation	339
School-Home Connection	340
Lesson 1 Creating Multiplication Tables	341
Lesson 2 Multiplication and Division	343
Lesson 3 Writing Multiplication and Division Fact Families	345
Lesson 4 Connecting Pictures, Number Sentences, and Stories	347
Lesson 5 Problem Solving Strategy: *Act It Out*	349
Problem Solving Test Prep	350
Chapter Review/Assessment	351

Glossary 353

Name _____

Chapter 1

Counting Strategies
Paper Clip Patterns

You need
- large and small paper clips

Make patterns with paper clips.

STEP 1 Creating Patterns

How many different patterns did you make? _____

Draw one of your patterns here.

_____

STEP 2 Describing Patterns

Use words to tell about the pattern you drew.

_____

Which is your favorite pattern? Tell about it. _____

_____

STEP 3 Extending Patterns

Continue one of your patterns. How did you know what to do?

_____

_____

_____

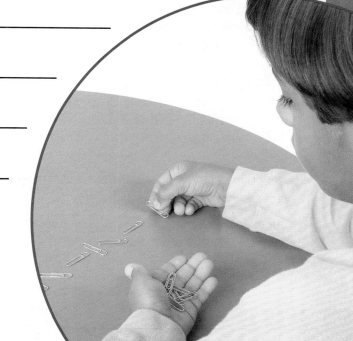

Investigation

School-Home Connection

Dear Family,
Today we started Chapter 1 of *Think Math!* In this chapter, I will explore numbers, number lines, patterns, skip-counting, addition, subtraction, and even multiplication. There are NOTES on the Lesson Activity Book pages to explain what I am learning every day.

Here are some activities for us to do together at home. These activities will help me understand numbers and counting patterns.

Love,

_____

Family Fun

What's My Number?

Work with your child to play a game called *What's My Number?* Your child will play this game later in this chapter.

- Tell your child you are thinking of a number from 1 to 8.
- Your child asks up to four *yes/no* questions to find the secret number. Each question should get rid of several numbers at once. Some good questions to ask are: "Is your number odd?" or "Is your number less than 5?"
- After each question, your child crosses off the numbers that have been eliminated.
- Your child wins the game if he or she guesses the secret number with up to 4 questions.

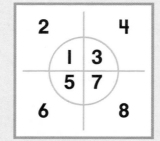

Number Puzzle

Work with your child to complete the number puzzle.

Across

1. $7 + 7 =$ ___
3. 170, 180, 190, ___
4. $1 + 2 + 3 + 4 + 5 =$ ___
6. 15, 17, 19, ___

Down

1. 8, 10, ___, 14
2. 100, 200, 300, ___
5. 40, 45, 50, ___
7. $8 + 8 =$ ___

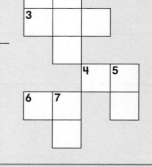

Chapter 1 Lesson 1

Repeating and Growing Patterns

NCTM Standards 2, 6, 7, 8, 9, 10

What comes next? Continue each pattern.

1.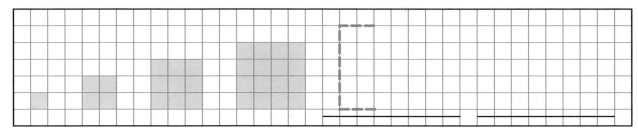

2. X O X O X O X O X O X ___ ___ ___

3. 4 5 6 4 5 6 4 5 6 ___ ___ ___

4.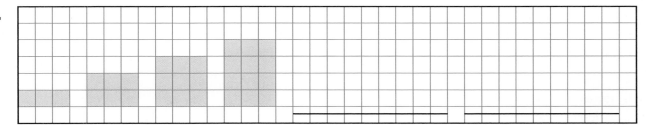

5. Make your own pattern. Draw it here.

NOTE: Your child is learning about patterns that repeat and grow. Together, look around your home for different patterns.

   III three 3

**Is it a repeating pattern? Circle *yes* or *no*.
If *yes*, circle the pattern unit.**

6.    yes   no

7.    yes   no

8. **5   10   15   20   25**   yes   no

9.    yes   no

 10. Choose a pattern from above that does NOT repeat. How does the pattern grow?

_____

_____

_____

Problem Solving

11. Carey gets 5 cents each day. In how many days will she have 25 cents? Use words, numbers, or pictures to explain.

_____ days

Chapter 1 Lesson 2

Working with Number Patterns

NCTM Standards 1, 2, 6, 9, 10

1. Continue the pattern. What is missing?

Stair-Step Table	
Number of Steps Up	Number of Dots
1	1
2	3

Hi! We are your Math Pals.

We are here to help!

How many dots are on each card?

2.

3.

4.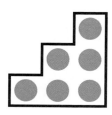

5. 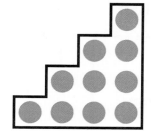

NOTE: Your child is using Stair-Step Cards to show patterns. The cards can be put together to make squares and show designs.

 I V five 5

Each figure is made from two cards. How many dots are in each figure? Record below.

6.   7. 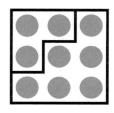   8.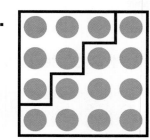

9. 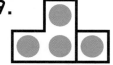   10. 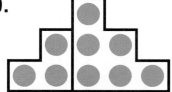   11.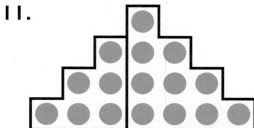

12. 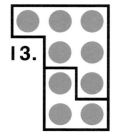   13. 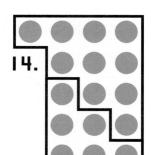   14. (shown with 13)   15. 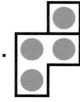   16.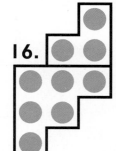

6	7	8		9	10	11		12	13	14		15	16
4													

Challenge

How many dots are missing from each figure?

17.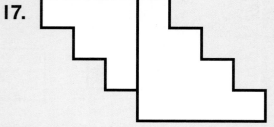

18.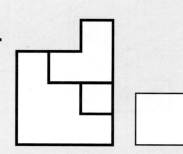

Chapter 1 Lesson 3

Writing Number Sentences

NCTM Standards 1, 2, 6, 7, 8, 9, 10

Name _____ Date _____

Write number sentences to go with each figure.

1.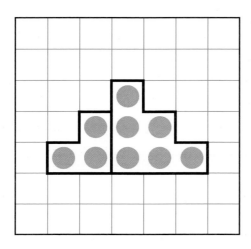

    $9 - 3 = 6$

    $1 + 2 + 3 + 2 + 1 = 9$

    $5 + 3 + 1 = 9$

2.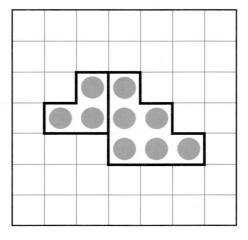

3.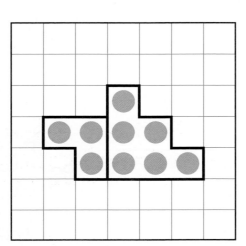

NOTE: Each figure is made from two Stair-Step Cards. Your child is learning to write number sentences about the dots in each figure by looking at the cards, the rows, and the columns.

VII seven 7

Write number sentences to go with each figure.

4.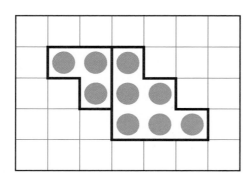

5.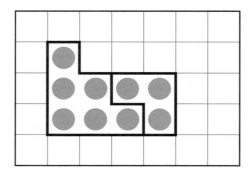

6.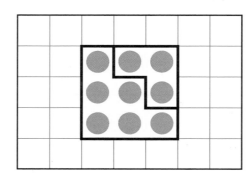

Problem Solving

7. How can both 3 + 6 = 9 and 6 + 3 = 9 tell about the same picture?

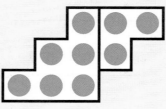

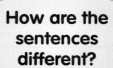

How are the sentences different?

Chapter 1 Lesson 4: Adding and Subtracting on the Number Line

NCTM Standards 1, 2, 6, 9, 10

Draw the jump.

What is missing?

1.

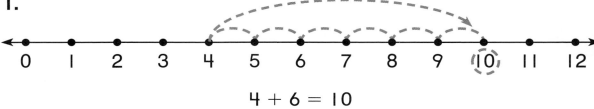

$4 + 6 = 10$

2.

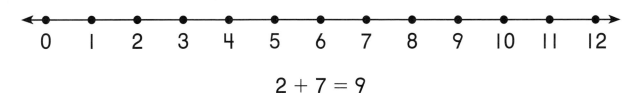

$2 + 7 = 9$

3.

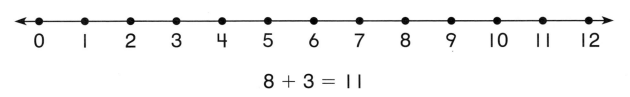

$8 + 3 = 11$

4.

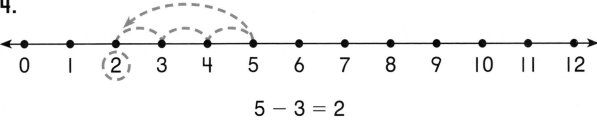

$5 - 3 = 2$

5.

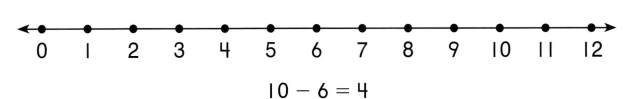

$10 - 6 = 4$

NOTE: Your child is using jumps on the number lines to show addition and subtraction.

IX nine 9

What number sentence is shown by the jump?

6.
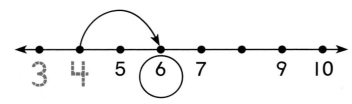
4 + 2 = 6

7.
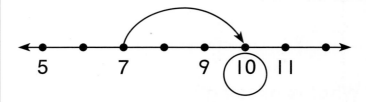
☐ + ☐ = ☐

8.
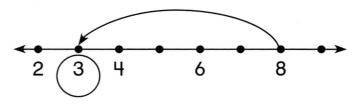

☐ − ☐ = ☐

9.
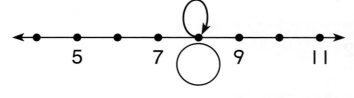
☐ + ☐ = ☐

10.
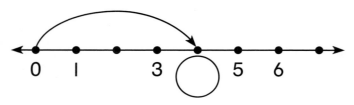

☐ + ☐ = ☐

11.
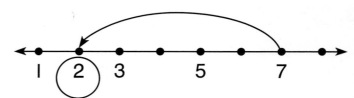

☐ − ☐ = ☐

Challenge
Make your own.

12.
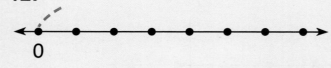
0 + ☐ = ☐

13.
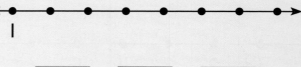
☐ − ☐ = ☐

Name _____ Date _____

Chapter 1 Lesson 5

Completing Number Sentences

NCTM Standards 1, 2, 6, 10

What number is missing?

1.
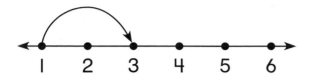

| 1 | + | 2 | = | 3 |

2.
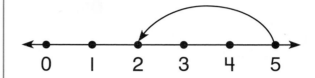

| 5 | − | 3 | = | |

3.
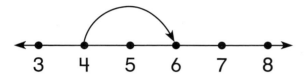

| | + | 2 | = | 6 |

4.
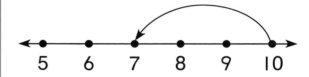

| | − | 3 | = | 7 |

5.
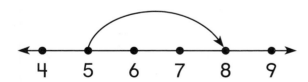

| 5 | + | | = | 8 |

6.
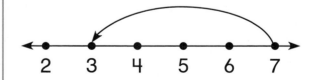

| 7 | − | | = | 3 |

NOTE: Your child is learning to relate jumps on a number line to number sentences. Each number sentence shows the starting number, the jump size, and the landing number.

   XI eleven

What number is missing?

7.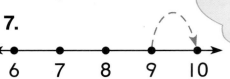

   Now draw the jump.

   | 9 | + | | = | 10 |

8.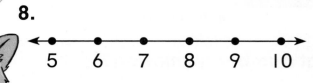

   | 10 | − | | = | 8 |

9.

   | | + | 3 | = | 10 |

10.

   | | − | 4 | = | 6 |

Challenge

11. Find as many ways as you can.

	+		=	10
	+		=	10
	+		=	10
	+		=	10
	+		=	10
	+		=	10

	+		=	10
	+		=	10
	+		=	10
	+		=	10
	+		=	10
	+		=	10

12 twelve XII

Chapter 1 Lesson 6: Skip-Counting on the Number Line

NCTM Standards 1, 2, 5, 6, 7, 8, 9, 10

Skip-count. What is missing?

1. Start at 0. The jump size is 2.

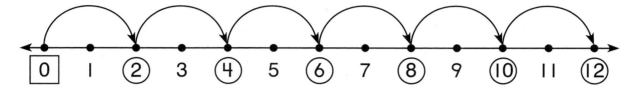

Number of Jumps	0	1	2	3	4	5	6	7
Landing Number	0	2	4	6				

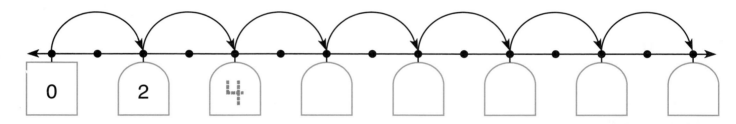

2. Start at 1. The jump size is 2.

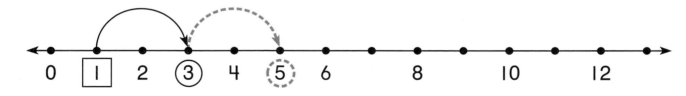

Number of Jumps	0	1	2	3	4	5	6	7
Landing Number	1	3	5					

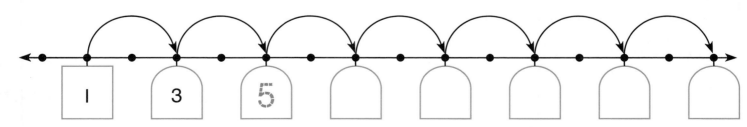

NOTE: Your child is learning to skip-count by different numbers on the number line. Together, practice skip-counting by twos from different starting numbers.

XIII   thirteen   13

Skip-count. What are the missing numbers?

3.
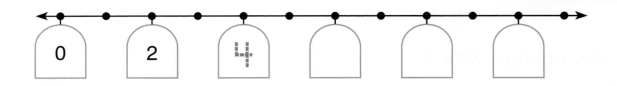

4.
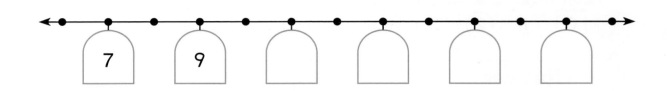

5.
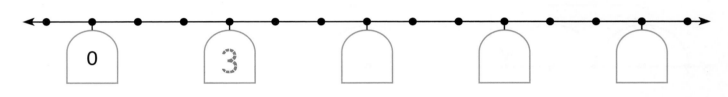

6. Make your own.

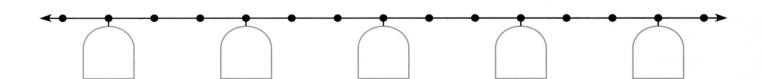

Problem Solving

7. Gracie says she can start at 0 and skip-count to 12 by threes. Tal says he can start at 0 and skip-count to 12 by fours. Who is right? Explain.

_____

_____

Chapter 1
Lesson 7
More Skip-Counting on the Number Line
NCTM Standards 1, 2, 6, 7, 8, 9, 10

What is missing?

1.

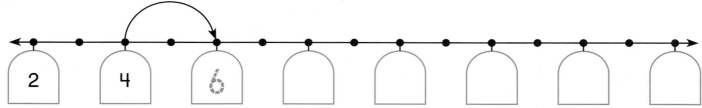

| 4 | + | | = | 6 |

2.

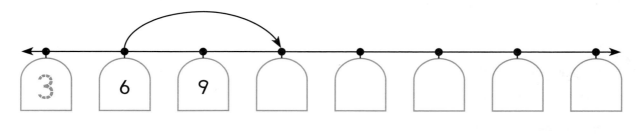

| 6 | + | | = | |

3.

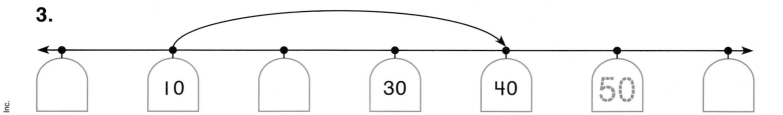

| | + | | = | |

NOTE: Your child is learning to work with number lines where the space between the dots is worth more than 1. The number lines can be used to help complete the addition sentences.

 3   XV fifteen   15

4. What is missing?

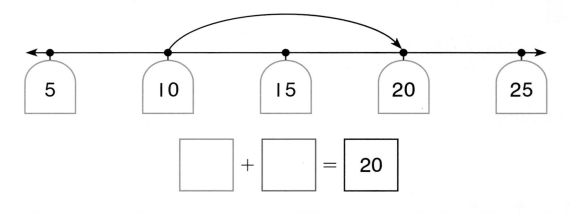

5. Make your own number line to show 20 + 8 = 28.

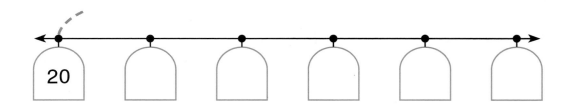

Problem Solving

6. You want to show 40 + 10 = 50. What are some different ways to label the number line? Explain.

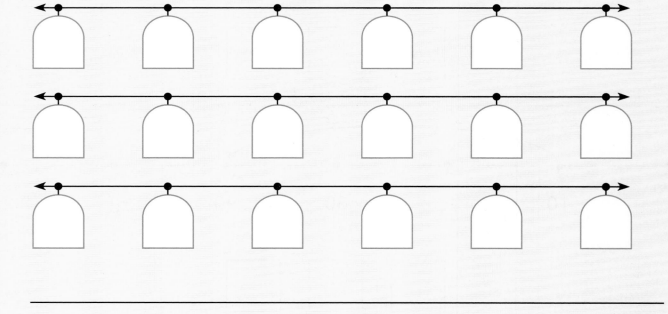

Chapter 1, Lesson 8: Systematic Counting

NCTM Standards 1, 2, 6, 9, 10

How many different towers can you build?
Follow the rules. Color to show the towers.
Mark an X on towers you do not color.

Rules: • Use one blue cube in each tower.
• Use cubes of another color to make the right height.

Height	Different Towers	Number of Towers
1. 1 cube tall		
2. 2 cubes tall		
3. 3 cubes tall		

NOTE: Your child is using cubes to build all possible towers for each height. Only one of the cubes is blue and the rest are another color.

20 − 3   XVII   seventeen   17

How many different towers can you build?

Height	Different Towers	Number of Towers
4. 4 cubes tall		_____
5. 5 cubes tall		_____
6. 6 cubes tall		_____

Challenge

7. What is missing?

Number of Cubes	1	2	3					
Number of Different Towers								

18 eighteen XVIII 18 9 + 9

Chapter 1
Lesson 9

Finding Ways to Make 10

NCTM Standards 1, 2, 6, 7, 8, 9, 10

Sums of 10 Search

1. Which pairs make 10? Circle them as fast as you can.

4/6	6/4	6/5	8/2	9/1	1/8	5/6	2/8
3/6	3/7	8/2	9/1	6/3	2/9	2/8	4/6	5/5
6/3	7/3	2/8	2/9	7/3	7/2	8/2	9/2	3/6
4/6	5/6	5/5	3/7	4/6	5/5	6/4	7/3	8/2
9/1	2/9	7/2	7/3	8/2	3/8	4/7	5/5	1/8
0/9	8/2	6/4	7/3	4/6	9/1	6/3	1/9	3/7

NOTE: Your child is learning to quickly recognize pairs of numbers with a sum of 10. You can practice by saying a number and having your child name that number's partner to make 10.

2. How many ways can you make 10?

Cover some, all, or no dots.

Uncovered	Covered
3	7

Problem Solving

3. I am thinking of two numbers with a sum of 10. One of the numbers is even. What can you say about the other number? What two numbers might they be?

_____

_____

Chapter 1
Lesson 10: Previewing Multiplication, Part I

NCTM Standards 1, 2, 3, 6, 9, 10

How many intersections are there? Write the missing numbers.

An intersection is where two lines meet.

1.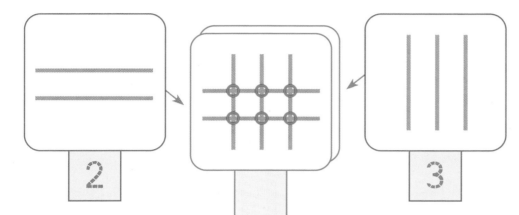
   2   3

2.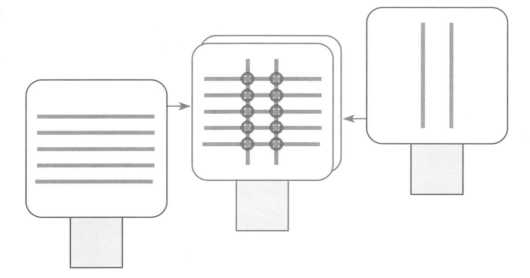

NOTE: Your child is exploring and counting intersecting lines. Pictures like these will be used later to model multiplication.

20 + 1 △ XXI twenty-one 21

What is missing?

3.

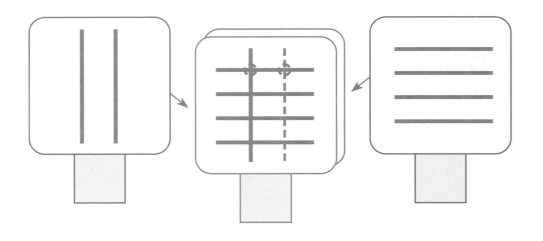

4.

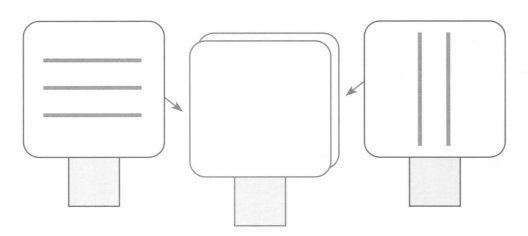

Challenge

5. Make your own.

What will you do first?

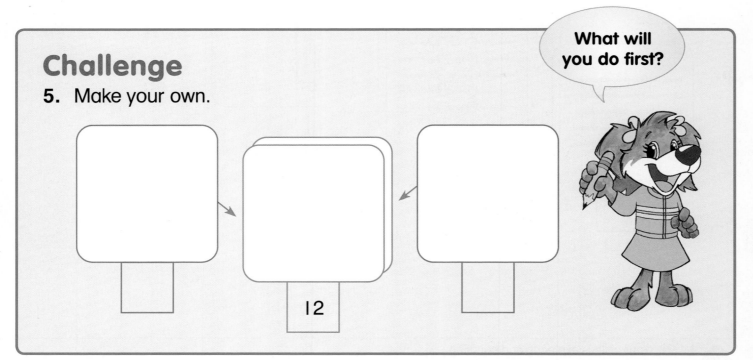

Chapter 1
Lesson 11: Previewing Multiplication, Part II

NCTM Standards 1, 2, 3, 6, 9, 10

What words can you make?

1.
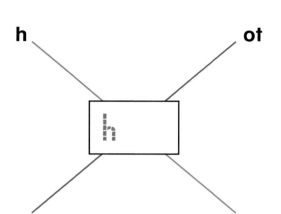

2.
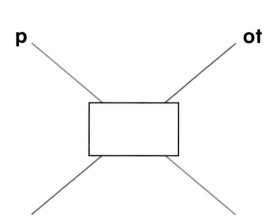

3.
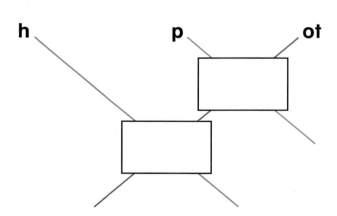

4.
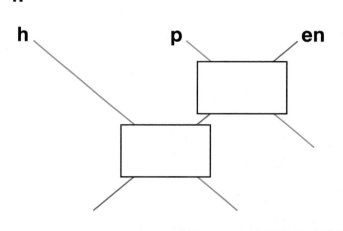

5.
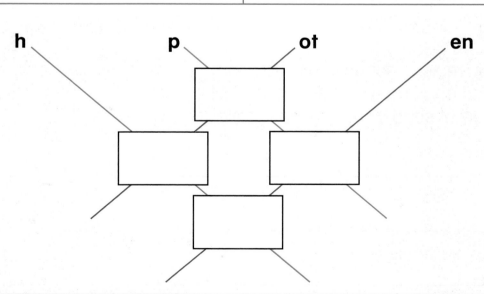

NOTE: Your child is learning to make words by combining letters at the intersections of lines. Pictures like these will be used later to model multiplication.

20 + 3   XXIII   twenty-three   23

6. What words can you make?

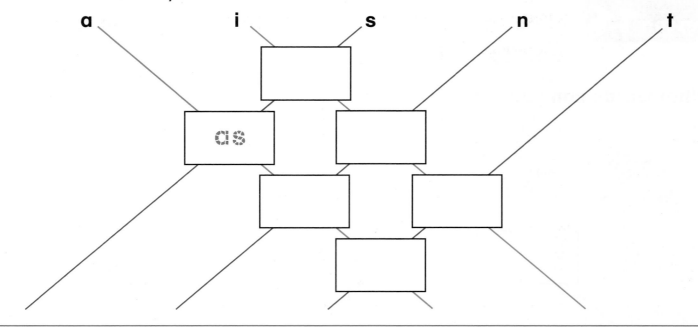

a   i   s   n   t

as

7. Make your own.

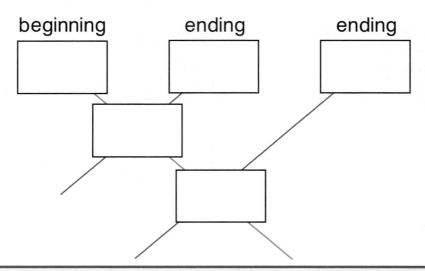

beginning   ending   ending

Problem Solving

8. Kermit is making sandwiches with one meat and one cheese. Write a list of all the different sandwiches he can make.

Meat	Cheese
bologna	American
turkey	Swiss

_____ _____

_____ _____

_____ _____

24   twenty-four   XXIV      2 dozen

Chapter 1
Lesson 12

Problem Solving Strategy
Look for a Pattern

NCTM Standards 1, 2, 6, 7, 8, 9, 10

Understand
Plan
Solve
Check

1. What is the number of the third house after 18 South Street? Explain.

_____

_____

_____

2. What is the number of the house where the next tree should be planted? Explain.

_____

_____

_____

3. Tammy is building a fence around her yard. Draw the next two fence posts. Explain how you know what to draw.

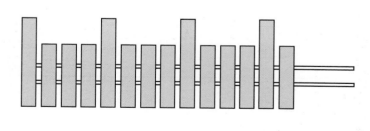

_____

_____

_____

NOTE: Your child is exploring different ways to solve problems. Sometimes looking for a pattern is an efficient way to solve a problem.

Problem Solving Test Prep

1. Carla has 6 games. Jeff has 4 games. How many games do they have altogether?

    Ⓐ 2 games

    Ⓑ 6 games

    Ⓒ 10 games

    Ⓓ 24 games

2. There are 11 children on the playground. 8 are on the swings. The rest are playing catch. Which number sentence shows how many children are playing catch?

    Ⓐ $11 + 8 = 19$

    Ⓑ $11 - 8 = 3$

    Ⓒ $8 + 8 = 16$

    Ⓓ $11 - 4 = 7$

✏️ Show What You Know

3. Matt bakes 15 muffins. He gives some to his mother. He gives 3 fewer muffins to his brother. He has 6 muffins left. How many muffins did he give to his mother?

    _____ muffins

    Explain how you found the answer.

4. The highest temperature on Monday was 47°. The temperature went up 2° each day. On what day will the highest temperature be 53°?

    Explain how you found the answer.

Name _____ Date _____

Chapter 1 Review/Assessment
NCTM Standards 1, 2, 3, 6, 7, 8, 9, 10

1. Continue the pattern. Lesson 1

2. How many dots are in each figure? Lesson 2

 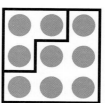 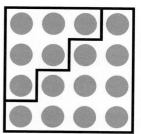

_____ _____ _____

Write number sentences to go with each figure. Lesson 3

3.

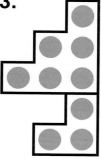

4.

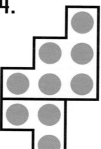

What number sentence is shown by the jump? Lesson 4

5.

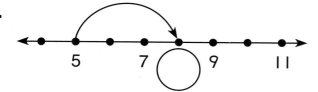

6.

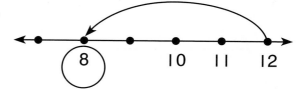

What number is missing? Lesson 5

7.

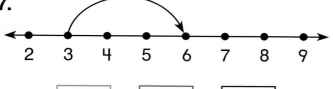

☐ 3 + ☐ 3 = ☐

8.

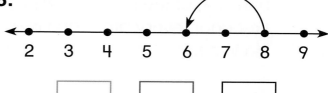

☐ − 2 = ☐ 6

What is missing? Lessons 6, 7

9.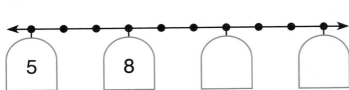
5 8

10.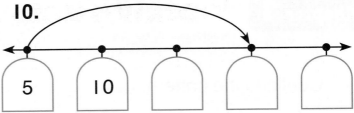
5 10

5 + ☐ = ☐

11. Circle pairs with a sum of 10. Lesson 9

| 8 | 6 | 1 | 3 | 5 | 4 | 3 | 2 | 9 | 5 |
| 2 | 5 | 9 | 7 | 5 | 7 | 6 | 8 | 1 | 5 |

12. How many intersections are there? Lesson 10

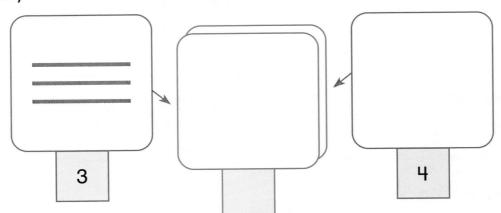

3   4

Problem Solving Lesson 12

13. At Mt. Way School, there are 7 doors on one side of the hallway. Starting with the first door, every other one is painted blue. How many blue doors are there?

_____ blue doors

28 twenty-eight XXVIII △ 14 + 14

Chapter 2

Name _____

Working with 10
Stories About 10

You need
- counters
- drawing paper

Take 10 counters. Separate them into two sets.

STEP 1 Counting Sets

How many counters do you have in all? _____

Does the total change when you move them around? _____

Explain your answer. _____

_____

_____

STEP 2 Telling Stories

Write a number story about your two sets.
Draw a picture to show the story.

_____

_____

STEP 3 Recording Sentences

What number sentence can describe the sets?

_____

What other number sentence can describe the same sets?

_____

Investigation

School-Home Connection

Dear Family,
   Today we started Chapter 2 of *Think Math!* In this chapter, I will work with number pairs that have a sum of 10 (such as $4 + 6$). I will also start at any number and add or subtract 10 and develop strategies to look at a number and tell how much more or less than 10 it is. There are NOTES on the Lesson Activity Book pages to explain what I am learning every day.
   Here are some activities for us to do together at home. These activities will help me practice working with 10.

Love,

_____

Family Fun

Fingers Up for 10!

Work with your child to add two numbers by making 10.

- Name two numbers from 5 to 10. Show one number with your fingers. Your child shows the other number. Make sure to use all of the fingers of one hand for the first 5 of each number.

- Add the two numbers by making 10. Put your hand of 5 with your child's hand of 5. This makes 10. Then add the rest of the fingers and add it to 10.

- Repeat several times with different pairs of numbers from 5 to 10.

X-Concentration

Work with your child to practice adding two numbers with a sum of 10.

- Gather a deck of playing cards and remove the 10s, jacks, queens, and kings.
- Mix the 36 cards. Place them face down in 4 rows of 9 like this.

- Take turns with your child. On each turn, a player turns two cards face up. If their sum is 10, the player takes the cards and goes again. The player continues until the pair does not add up to 10.
- Take turns until all cards are paired.
- As a variation, play with cards from only 2 suits.

Finding Sums of 10

Chapter 2, Lesson 1

NCTM Standards 1, 2, 6, 7, 8, 9, 10

1. Find different ways to put 10 counters in two sets. Record below.

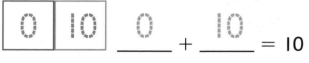

0 + 10 = 10

____ + ____ = 10

____ + ____ = 10

____ + ____ = 10

____ + ____ = 10

____ + ____ = 10

____ + ____ = 10

____ + ____ = 10

Use the same 10 counters each time.

____ + ____ = 10

____ + ____ = 10

____ + ____ = 10

____ + ____ = 10

NOTE: Your child is learning all of the different addition sentences with a sum of 10.

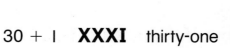

What is missing?

2.

_____ + _____ = _____

3.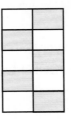

_____ + _____ = _____

4.

_____ + _____ = _____

5.

_____ + _____ = _____

Challenge

What is the value of in each problem?

6.

_____ + 2 = 10

7.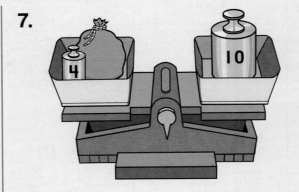

_____ + 4 = 10

Chapter 2 Lesson 2: Introducing > and <

NCTM Standards 1, 2, 6, 7, 8, 9, 10

Write >, <, or =. Make each sentence true.

1. 1 + 4 ◯>◯ 1 + 3
2. 5 + 1 ◯ 5 + 2
3. 2 + 8 ◯ 1 + 7
4. 5 + 6 ◯ 6 + 5
5. 1 + 8 ◯ 1 + 6
6. 3 + 6 ◯ 2 + 6
7. 2 + 3 ◯ 3 + 2
8. 7 + 3 ◯ 3 + 7
9. 4 + 2 ◯ 2 + 8
10. 6 + 5 ◯ 5 + 5
11. 4 + 5 ◯ 1 + 3
12. 6 + 5 ◯ 7 + 5

NOTE: Your child is learning to compare addition expressions using the symbols <, >, or = to make each sentence true. < means "is less than", > means "is greater than", and = means "is equal to."

11 + 11 + 11   33   XXXIII   thirty-three   33

What is missing? Make each sentence true.

13. $4 + 5 < 4 + \underline{}$

14. $8 + 2 < 8 + \underline{}$

15. $3 + 2 > \underline{} + 2$

16. $5 + 3 > \underline{} + 3$

17. $4 + 5 < \underline{} + 5$

18. $1 + 6 > 1 + \underline{}$

19. $2 + 7 > \underline{} + \underline{}$

20. $3 + 9 > \underline{} + \underline{}$

21. $7 + 8 < \underline{} + \underline{}$

22. $10 + 9 < \underline{} + \underline{}$

23. What addition expressions make the sentence true?

$7 + 4 > \underline{} + \underline{}$

$\underline{} + \underline{}$ $\underline{} + \underline{}$ $\underline{} + \underline{}$

$\underline{} + \underline{}$ $\underline{} + \underline{}$ $\underline{} + \underline{}$

Problem Solving

24. Ned has two stools. The metal stool is 5 inches tall. The wooden stool is 14 inches tall. On which stool can Ned stand higher? Explain.

Chapter 2 Lesson 3: Finding and Comparing Sums to 10

NCTM Standards 1, 2, 8, 9, 10

Sums Greater than 10 Search

1. Which pairs make sums greater than 10? Circle them as fast as you can.

5/5	5/6	5/7	4/4	4/6	7/4	7/3	4/7	2/8
3/7	3/5	3/8	8/1	6/4	6/7	7/2	6/5	4/6
4/7	2/7	7/3	9/0	1/9	2/9	1/8	3/6	6/5
0/10	7/5	9/1	4/5	0/9	8/3	8/2	9/2	10/0

NOTE: Your child is learning to identify number pairs with sums greater than 10 and compare addition problems to 10 or 20.

Write >, <, or =.

2. 10 ⊘< 7 + 5

3. 20 ◯ 17 + 5

4. 2 + 9 ◯ 10

5. 12 + 9 ◯ 20

6. 8 + 4 ◯ 10

7. 8 + 14 ◯ 20

8. 10 ◯ 3 + 7

9. 20 ◯ 3 + 17

10. 5 + 6 ◯ 2 + 8

11. 15 + 6 ◯ 20

12. 2 + 7 ◯ 10

13. 2 + 17 ◯ 20

Challenge

Write >, <, or =.

14. If 6 + 7 ◯ 10, then 16 + 17 ◯ 30.

15. If 8 + 2 ◯ 10, then 18 + 12 ◯ 30.

Chapter 2 Lesson 4: Mastering Sums of 10

NCTM Standards 1, 2, 6, 7, 8, 9, 10

Write >, <, or =. Make each sentence true.

1. 4 + 6 ◯> 4 + 5
2. 9 + 2 ◯ 1 + 9
3. 1 + 8 ◯ 1 + 9
4. 3 + 6 ◯ 4 + 6
5. 4 + 6 ◯ 3 + 7
6. 1 + 8 ◯ 2 + 8
7. 3 + 7 ◯ 7 + 4
8. 4 + 6 ◯ 2 + 8
9. 7 + 3 ◯ 5 + 4
10. 1 + 3 ◯ 1 + 9
11. 8 + 3 ◯ 2 + 8
12. 5 + 5 ◯ 6 + 5

NOTE: Your child is memorizing the facts with a sum of 10. These facts will help him or her later to solve other facts.

What is missing? Make each sentence true.

13. 4 + 6 = 3 + __7__

14. 5 + 5 = 8 + ____

15. 1 + 9 = ____ + 8

16. 7 + 3 = ____ + 7

17. 5 + 5 < ____ + 5

18. 9 + 1 > 1 + ____

19. 6 + 4 < ____ + 9

20. 0 + 10 > 4 + ____

21. ____ + 4 < 3 + 7

22. ____ + 6 > 3 + 7

 23. Make the sentence true. Explain your answer.

2 + 8 < 2 + ____

_____

_____

_____

Challenge

24. How can you make four number pairs with sums of 10? Draw lines to match the numbers below.

1   2   3   4   5   6   7   8   9

Chapter 2 Lesson 5: Adding and Subtracting with 10

NCTM Standards 1, 2, 6, 7, 8, 9, 10

Add 10. What number is missing?

1. 51 + 10 = 61
2. 4 + 10 = ☐
3. 32 + 10 = ☐
4. ☐ + 10 = 79
5. 28 + 10 = ☐
6. ☐ + 10 = 26

Move up exactly one row.

90	91	92	93	94	95	96	97	98	99
80	81	82	83	84	85	86	87	88	89
70	71	72	73	74	75	76	77	78	79
60	61	62	63	64	65	66	67	68	69
50	51	52	53	54	55	56	57	58	59
40	41	42	43	44	45	46	47	48	49
30	31	32	33	34	35	36	37	38	39
20	21	22	23	24	25	26	27	28	29
10	11	12	13	14	15	16	17	18	19
0	1	2	3	4	5	6	7	8	9

7.

| 20 | ☐ | ☐ | 77 | ☐ | ☐ | 95 | ☐ | ☐ |
| 10 | 20 | 24 | 67 | 58 | 82 | 85 | 90 | 46 |

8. 85 + 10 = ☐
9. ☐ + 10 = 20
10. 90 + 10 = ☐
11. ☐ + 10 = 56
12. 20 + 10 = ☐
13. 24 + 10 = ☐
14. 82 + 10 = ☐
15. 58 + 10 = ☐
16. ☐ + 10 = 77

 NOTE: Your child is learning to use a hundred grid to add and subtract 10 from any two-digit number.

40 − 1   XXXIX   thirty-nine   39

Subtract 10. What number is missing?

17. 72 − 10 = 62

18. 30 − 10 = ☐

19. 91 − 10 = ☐

20. ☐ − 10 = 33

21. ☐ − 10 = 14

22. 85 − 10 = ☐

Move down exactly one row.

90	91	92	93	94	95	96	97	98	99
80	81	82	83	84	85	86	87	88	89
70	71	72	73	74	75	76	77	78	79
60	61	62	63	64	65	66	67	68	69
50	51	52	53	54	55	56	57	58	59
40	41	42	43	44	45	46	47	48	49
30	31	32	33	34	35	36	37	38	39
20	21	22	23	24	25	26	27	28	29
10	11	12	13	14	15	16	17	18	19
0	1	2	3	4	5	6	7	8	9

23.

24	29	72	85	17	43	30	91	90
14		62				20		

Add and subtract.

24. 64 + 10 = ☐
    64 − 10 = ☐

25. 56 + 10 = ☐
    ☐ − 10 = 46

26. ☐ + 10 = 48
    ☐ − 10 = 28

Challenge
What is the missing number?

27.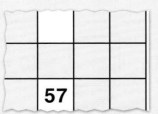

28.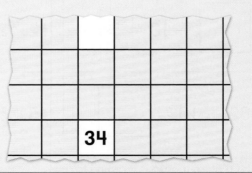

 forty XL 8

Chapter 2
Lesson 6: Finding How Close to 10

NCTM Standards 1, 2, 6, 7, 8, 9, 10

How close to 10 is the sum?

1. 7 + 1
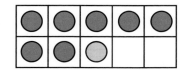
10 [−2]

2. 4 + 7
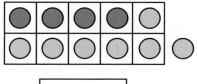
10 [ ]

3. 6 + 6
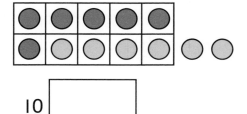
10 [ ]

4. 5 + 4
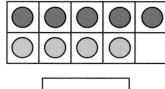
10 [ ]

5. 8 + 2
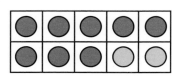
10 [ ]

6. 3 + 5
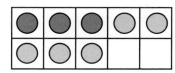
10 [ ]

7. 3 + 9
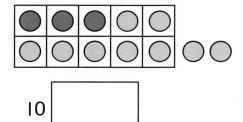
10 [ ]

8. 1 + 9
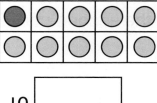
10 [ ]

NOTE: Your child is learning to identify how close a sum is to 10. Ask him or her to explain the exercises on this page.

40 + 1   **XLI**   forty-one   41

Make each sentence true.

9. 8 + 3 = 10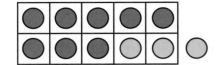

10. 2 + 7 = 10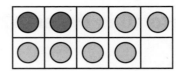

11. 4 + 4 = 10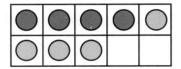

12. 4 + 8 = 10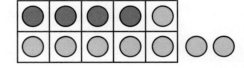

13. 9 + 3 = 10 ☐

14. 8 + 1 = 10 ☐

Make your own.

15. ☐ + ☐ = 10 + 1

16. ☐ + ☐ = 10 − 2

Problem Solving

17. Jill buys two packs of 10 hot dogs each. She serves 6 hot dogs on Friday and 6 on Saturday. How many hot dogs does she use from the second pack? Explain.

_____

_____

42 forty-two XLII 21 + 21

Chapter 2 Lesson 7

Adding Numbers by Making 10

NCTM Standards 1, 2, 6, 7, 8, 9, 10

Rewrite each fact with a 10.

Make 10 with 5 from each number. Then add what is left over.

1. $6 + 8 =$ 14

   $10 + 4 = 14$

2. $7 + 9 =$ ☐

   10 _____

3. $8 + 5 =$ ☐

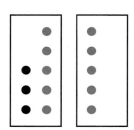

_____

4. $9 + 8 =$ ☐

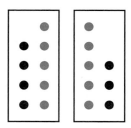

_____

5. $8 + 7 =$ ☐

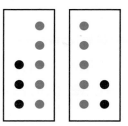

_____

6. $9 + 9 =$ ☐

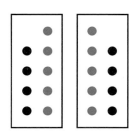

_____

7. $7 + 6 =$ ☐

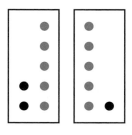

_____

8. $6 + 9 =$ ☐

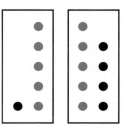

_____

NOTE: Your child is learning a strategy to add two numbers by making 10 with a 5 from each number and then adding the remaining numbers to 10.

$40 + 3$ **XLIII** forty-three

Add fives to find each sum. What is missing?

9. 6 + 9 =

5 + _1_ = 6
5 + _4_ = 9
10 + _5_ = ☐15☐

10. 8 + 7 =

5 + ___ = 8
5 + ___ = 7
10 + ___ = ☐

11. 9 + 5 =

5 + ___ = 9
5 + ___ = 5
10 + ___ = ☐

12. 7 + 6 =

___ + ___ = 7
___ + ___ = 6
___ + ___ = ☐

13. 8 + 9 =

___ + ___ = 8
___ + ___ = 9
___ + ___ = ☐

14. 6 + 8 =

___ + ___ = 6
___ + ___ = 8
___ + ___ = ☐

 15. How many different ways can you solve 7 + 9? Use words, numbers, or pictures to explain.

Problem Solving

16. Kat made 10 to solve 9 + 6. She did not add fives. What strategy might she have used?

_____

_____

44   forty-four   XLIV   11 + 11 + 11 + 11

Chapter 2
Lesson 8

Rounding to the Nearest 10

NCTM Standards 1, 2, 6, 7, 8, 9, 10

Name _____ Date _____

Write the tens that sandwich the number on each number line. Which is the nearest ten?

1.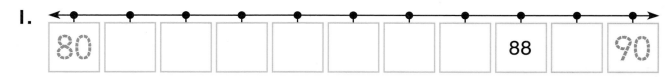

   The nearest ten to 88 is __90__.

2.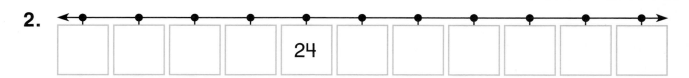

   The nearest ten to 24 is _____.

3.

   The nearest ten to 35 is _____.

4.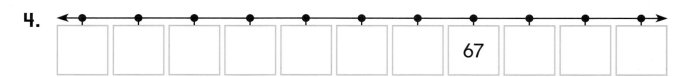

   The nearest ten to 67 is _____.

 5. What is the rule for rounding to the nearest ten? _____

_____

_____

NOTE: Your child is learning a strategy to round two-digit numbers to the nearest multiple of ten. This skill may be used in various ways throughout mathematics.

 9 **XLV** forty-five 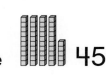 45

Make a sandwich with each number line. What number might come between the tens?

Write the number in position on the number line.

6. The nearest ten is 30.

7. The nearest ten is 50.

8. The nearest ten is 80.

9. The nearest ten is 40. Write all possible numbers that could be rounded to 40.

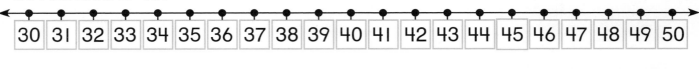

_____   _____   _____   _____   _____

_____   _____   _____   _____   _____

Problem Solving

10. Daunte wants to buy an apple for 70¢. He has 67¢. He says he has enough money because 67 rounds to 70. Is he right? Use words, numbers, or pictures to explain.

_____

_____

_____

Chapter 2, Lesson 9

Problem Solving Strategy
Solve a Simpler Problem

NCTM Standards 1, 2, 4, 6, 7, 8, 9, 10

Understand
Plan
Solve
Check

1. Karen made 2 dozen cookies. Jim made 4 fewer cookies than Karen. How many cookies did they make altogether?

   How did you find the answer?

   _____ cookies

2. Karen gave 8 cookies to Sasha and 5 to her brother. Jim gave 9 cookies to Oleg and 6 to his sister. Who gave away more cookies, Karen or Jim?

   How did you find the answer?

   _____

3. There are 53 tiles on the floor. It takes 1 minute to paint each tile. How many minutes will it take to paint the entire floor?

   How did you find the answer?

   _____ minutes

NOTE: Your child is exploring different ways to solve problems. Sometimes solving a simpler problem is an efficient way to solve a problem.

50 − 3   **XLVII**   forty-seven   47

Problem Solving Test Prep

1. Sue practices her violin for 3 hours each day. In how many days will she practice for a total of 12 hours?

    Ⓐ 36 days

    Ⓑ 15 days

    Ⓒ 9 days

    Ⓓ 4 days

2. Cliff has 2 coins. Both coins are the same. Which amount could he NOT have as the total?

    Ⓐ 10¢

    Ⓑ 20¢

    Ⓒ 40¢

    Ⓓ 50¢

Show What You Know

3. Raj, Bev, and Al are running a race. One person comes in first place, one in second, and one in third place. How many different ways can they finish the race?

    _____ ways

    Explain how you found the answer.

    _____

    _____

    _____

    _____

    _____

    _____

4. Pia saw birds and dogs at the pet store. She saw 4 heads and 12 feet. How many of each animal did she see?

    _____ birds _____ dogs

    Explain how you know.

    _____

    _____

    _____

    _____

    _____

    _____

forty-eight   XLVIII   4 dozen

Chapter 2 Review/Assessment

NCTM Standards 1, 2, 6, 7, 8, 9, 10

What is missing? Lesson 1

1.

_____ + _____ = _____

2.

_____ + _____ = _____

Write >, <, or =. Lessons 2, 3

3. 1 + 5 ◯ 1 + 2

4. 10 + 6 ◯ 20

What is missing? Make each sentence true. Lesson 4

5. 8 + 2 < 3 + _____

6. _____ + 3 < 4 + 6

Add or subtract 10. What number is missing? Lesson 5

7. 68 + 10 = _____

8. _____ + 10 = 33

9. 75 − 10 = _____

10. _____ − 10 = 41

90	91	92	93	94	95	96	97	98	99
80	81	82	83	84	85	86	87	88	89
70	71	72	73	74	75	76	77	78	79
60	61	62	63	64	65	66	67	68	69
50	51	52	53	54	55	56	57	58	59
40	41	42	43	44	45	46	47	48	49
30	31	32	33	34	35	36	37	38	39
20	21	22	23	24	25	26	27	28	29
10	11	12	13	14	15	16	17	18	19
0	1	2	3	4	5	6	7	8	9

50 − 1   XLIX forty-nine

How close to 10 is the sum? Lesson 6

11. 3 + 8

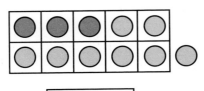

10 ☐

12. 6 + 4

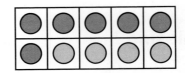

10 ☐

Rewrite each fact with a 10. Lesson 7

13. 9 + 6 = ?

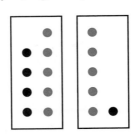

_____

14. 7 + 8 = ?

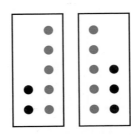

_____

15. 8 + 5 = ?

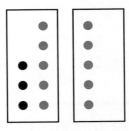

_____

16. What number might come between the tens?
The nearest ten is 70. Lesson 8

60 ———————————————————— 70

Problem Solving Lesson 9

17. Maya's grade at school has 90 students.
Her brother's grade has 70 students.
How many students are in the two grades together? _____ students

How did you find the answer?

Name _____

Chapter 3: Place Value
Counting Counters

You need
- zip-top bag of counters

Look at the counters in your group's bag.

STEP 1 Estimating

How many counters do you think are in your bag? _____

How did you decide? _____

_____

STEP 2 Counting

What is the exact number of counters in your bag? _____

How did you find the exact number? _____

_____

_____

STEP 3 Counting Another Way

What is another way to find the exact number? _____

_____

Which way do you like better? Explain.

_____

_____

_____

School-Home Connection

Dear Family,

Today we started Chapter 3 in *Think Math!* In this chapter, I will estimate, count, read and write numbers, and draw symbols for numbers to 999. There are NOTES on the Lesson Activity Book pages to explain what I am learning every day.

Here are some activities for us to do together at home. These activities will help me understand place value.

Love,

_____

Family Fun

Place-Value Match

Work with your child to prepare game cards to play *Place-Value Match*.

- Use index cards or slips of paper to make a set of 32 game cards. On 16 cards, write a three-digit number. On the remaining cards, write the number of hundreds, tens, and ones in each three-digit number.
- Partners shuffle the cards, and each one takes 6 cards. They put the remaining cards in a facedown stack.

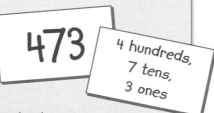

- Players take turns asking each other for cards to make a match. For example:

    Do you have a card for 4 hundreds, 7 tens, 3 ones?

- If the partner does not have the matching card, the player chooses a card from the stack. If the partner has the card, a match is made, and the player puts the pair of cards aside.
- The first player to match all of his or her cards wins the game.

Guess How Many?

Work with your child to estimate and count objects.

- Gather a container of small identical objects such as pennies, paper clips, or marbles. Ask your child to take a handful of the objects and count them.
- Without counting, each player guesses the number of objects in the container and writes the guess on a sheet of paper.
- Players then work together to group the objects into sets of ten to count the total number of objects. The objects are returned to the container.
- The player with the guess closer to the actual number is the winner.

Chapter 3 Lesson 1

Estimating and Counting Larger Numbers

NCTM Standards 1, 6, 7, 8, 9, 10

**How many are there? Use the set of 10 to estimate.
Then count to find the total.**

1.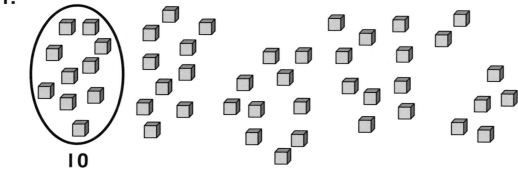

   45   47
   Estimate   Count

2.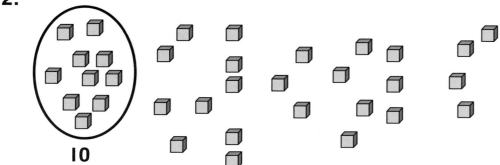

   _____   _____
   Estimate   Count

3.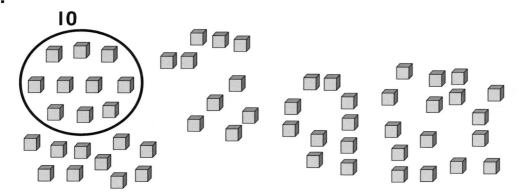

   _____   _____
   Estimate   Count

 NOTE: Your child is learning to estimate by using 10 as a benchmark and is making smaller sets to count a large collection of objects.

50 + 3   **LIII**   fifty-three   53

4. Draw as many squares as you can in the frame. Ask a classmate to estimate the total and make sets to count.

Estimate _____ Count _____

Challenge

5. How many are there? Estimate.
Count some and revise your estimate.
Count it all to find the total.

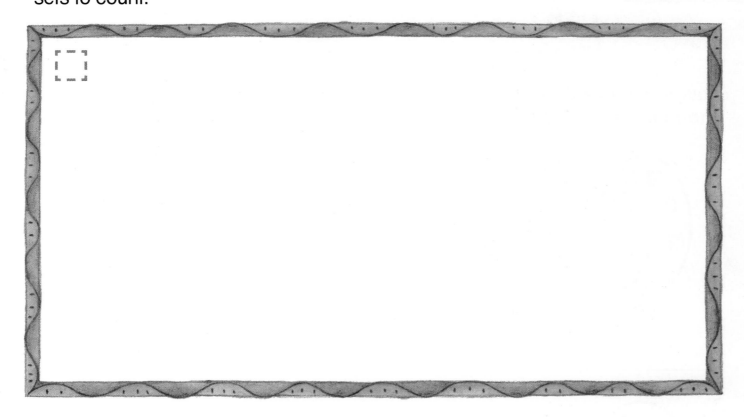

Estimate _____   Revised Estimate _____

Total _____

54 fifty-four LIV 27 + 27

Chapter 3 Lesson 2

Grouping by Tens and Hundreds

NCTM Standards 1, 6, 7, 8, 9, 10

Name _____ Date _____

Both pictures show the same number.
How many wheels do they show?

It is easier to count the wheels in sets of 10.

1.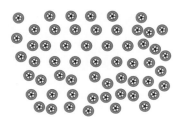

   [ ] ones

   5 tens   7 ones ⇒ 57

2.

   [ ] ones

   ___ tens   ___ ones ⇒ _____

3.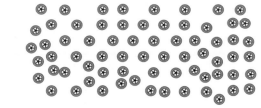

   [ ] ones

   ___ tens   ___ ones ⇒ _____

4.

   [ ] ones

   ___ tens   ___ ones ⇒ _____

NOTE: Your child is learning that it is sometimes easier to group objects by tens and hundreds when counting a large collection of objects.

    LV fifty-five 55

5. All three pictures show the same number.
   How many wheels do they show?

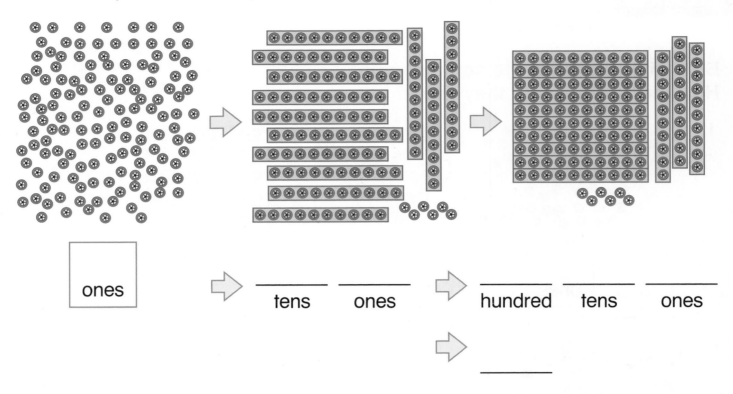

ones ⇒ ___ ___ ⇒ ___ ___ ___
        tens ones   hundred tens ones

                  ⇒ ___

---

6. Circle the set in Problem 5 that is easiest to count.
   Explain why you think this is so.

   _____

   _____

   _____

Problem Solving

7. A millipede is a small animal with many legs. This millipede has 115 pairs of legs. How many legs does the millipede have? (Hint: A pair is two.)

   ____ hundreds ____ tens ____ ones = ____ legs

Representing Two-Digit Numbers

Chapter 3
Lesson 3

NCTM Standards 1, 6, 8, 9, 10

What is missing? Draw symbols and write numbers.

1.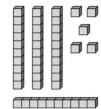

   __4__ tens __5__ ones = __40__ + __5__ = __45__

2.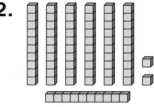

   ____ tens ____ ones = ____ + ____ = ____

3.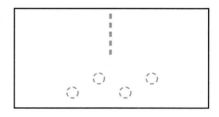

   ____ tens 4 ones = 10 + ____ = ____

4.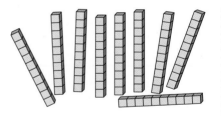

   ____ tens ____ ones = ____ + ____ = ____

5.

   5 tens 3 ones = ____ + ____ = ____

6. Make your own.

   ____ tens ____ ones = ____ + ____ = ____

NOTE: Your child is learning the value of the digits in two-digit numbers. Ask your child to tell how many tens and ones are in different numbers around you.

What number from the box will complete each riddle?

7. My tens digit has a value of 40.
   My ones digit is more than 5.

   _49_

15	19	22
28	32	35
43	49	53
69	72	85
88	90	94

8. My tens digit has a value of 90.
   My ones digit is less than 4.

   _____

9. Both of my digits are the same number. My tens digit has a value of 80.

   _____

10. My tens digit has a value less than 50. My ones digit has a value of 5.

    _____

11. My tens digit has a value less than 40. My ones digit is more than 7.

    _____

✏️ 12. Make up your own riddle about a number in the box.

_____

_____

_____

Problem Solving

13. I have more than 4 tens.
    I have between 6 and 9 ones.

    What number could I be? _____
    Draw a picture of the number.

58 fifty-eight LVIII 29 + 29

Chapter 3 Lesson 4: Representing Three-Digit Numbers

NCTM Standards 1, 6, 8, 9, 10

What is missing? Draw symbols and write numbers.

1.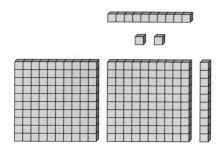

   __2__ hundreds __2__ tens __3__ ones

   __200__ + __20__ + __3__ = __223__

2.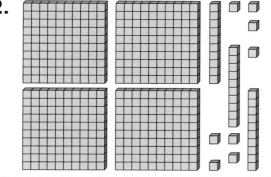

   _____ hundreds _____ tens _____ ones

   _____ + _____ + _____ = _____

3.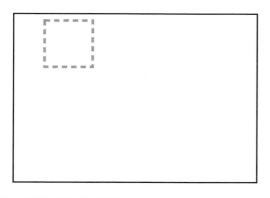

   _____ hundreds _____ tens _____ ones

   _____ + _____ + _____ = 706

4. Make your own.

   _____ hundreds _____ tens _____ ones

   _____ + _____ + _____ = _____

 NOTE: Your child is learning the value of digits in three-digit numbers. Ask your child to tell how many hundreds, tens, and ones are in different numbers you name.

60 − 1   **LIX**   fifty-nine   59

Toss 5 small counters onto the grid. Record the numbers. An example is given in Problem 5.

1	100	10	10	1	10	1
100	1	1	10	10	100	1
1	100	10	10	10	100	100

	Numbers That Counters Land On	**Total**
5.	10 + 100 + 1 + 1 + 10	122
6.	___ + ___ + ___ + ___ + ___	___
7.	___ + ___ + ___ + ___ + ___	___
8.	___ + ___ + ___ + ___ + ___	___

Challenge

What number is 10 more?

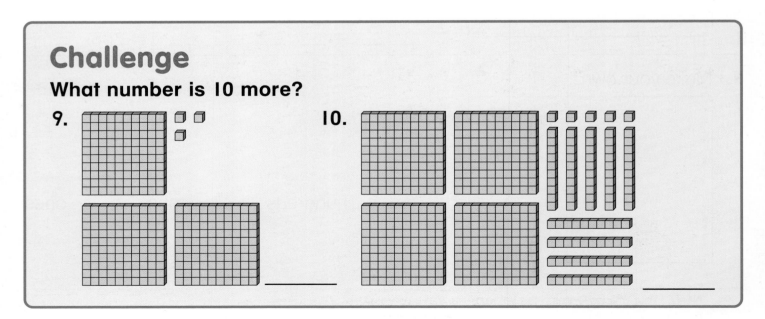

9.   10.

60   sixty   LX   12   5 dozen

Chapter 3 Lesson 5: Regrouping

NCTM Standards 1, 6, 7, 8, 9, 10

**Show each number with the fewest blocks.
Draw symbols for the blocks.**

1.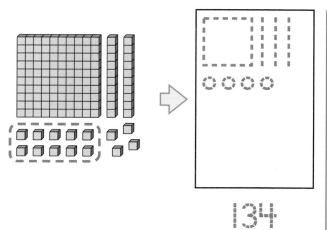

   134

2.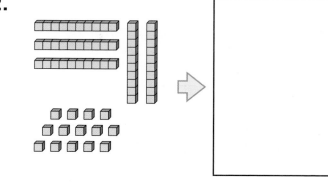

3.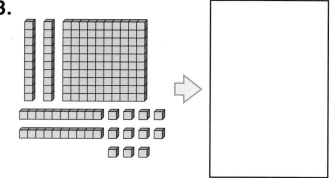

4.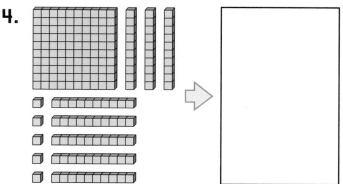

5.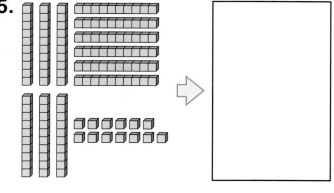

6.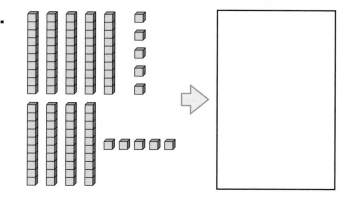

NOTE: Your child is learning to regroup tens and ones to show the same number in different ways.

60 + 1   LXI   sixty-one   61

Regroup in different ways.

7. | 24 |

___0___ tens ___24___ ones

___1___ tens ___14___ ones

_____ tens _____ ones

8. | 38 |

_____ tens _____ ones

_____ tens _____ ones

_____ tens _____ ones

_____ tens _____ ones

 9. Choose any two-digit number. How many ways can you regroup? Draw or write to show the ways.

Number _____

Problem Solving

10. Elva has 36¢. She has only dimes and pennies. What different combinations of coins might she have? Use words, numbers, or pictures to explain.

Using Place Value to Compare

Chapter 3 Lesson 6

NCTM Standards 1, 2, 6, 7, 8, 9, 10

Write the numbers. Then write >, <, or =.

1.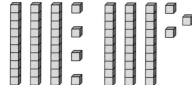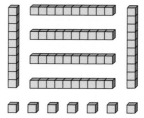

   67 ◯ ▢

2.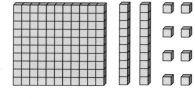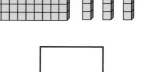

   ▢ ◯ ▢

3.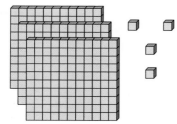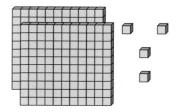

   ▢ ◯ ▢

4. Make your own.

   ▢ ◯ ▢

NOTE: Your child is learning to compare two- and three-digit numbers. Ask your child to compare the ages of various family members.

21 + 21 + 21   LXIII   sixty-three   63

Write a digit to make each sentence true.

5. 678 < __7__ 78

6. 364 < _____ 64

7. 420 > _____ 20

8. 7 _____ 8 < 768

9. 8 _____ 4 > 864

10. 2 _____ 7 < 267

11. 40 _____ > 403

12. 26 _____ < 269

13. 51 _____ < 515

14. 623 > 6 _____ 3

15. _____ 92 > 492

16. 134 < 13 _____

 17. Write all possible answers for Problem 16. Use words, numbers, or pictures to explain.

_____ _____ _____

_____ _____ _____

Problem Solving

18. Jim's soccer team got 32 goals this season. The number of goals last season was a lot less. It had the same ones digit. How many goals do you think they got last year? Explain.

Use what you know about place value to help you.

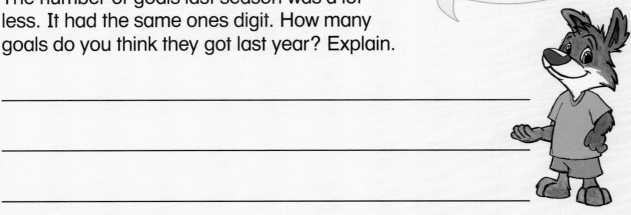

32 + 32

Chapter 3 Lesson 7
Connecting Numbers and Words
NCTM Standards 1, 2, 6, 7, 8, 9, 10

What number matches each word?

1. twenty-nine 29
2. twelve .... ____
3. sixty .... ____
4. one hundred ten .... ____
5. thirty-three .... ____
6. ninety-one .... ____
7. two hundred three .... ____
8. eighty-four .... ____

9. What is the order from smallest to biggest?

12   67   0   15   51   38

0 , ____ , ____ , ____ , ____ , ____

10. What is the order from biggest to smallest?

230   42   18   75   501   342   40   96

____ , ____ , ____ , ____ , ____ , ____ , ____ , ____

NOTE: Your child is learning to write numbers for number words and to order numbers. Ask your child to write the number for fifty-eight.

13 LXV sixty-five 65

Complete each number word in any way. Then write the numbers in order from biggest to smallest.

Use the list to help you write the words.

one	two	three
four	five	six
seven	eight	nine
twenty	thirty	forty
fifty	sixty	seventy
eighty	ninety	hundred

11. forty-___seven___ .... __47__

 ___thirty___-six .... __36__

 _____-four ..... _____

 seventy-_____ ... _____

 In order: _____,_____, _____, _____

12. _____ hundred fifty-five .... _____

 two hundred _____-two .... _____

 four hundred seventy-_____ .... _____

 _____ hundred twenty-_____ ...... _____

 In order: _____,_____, _____, _____

Challenge

13. What is the biggest number you can make from these digits? What is the smallest? Explain.

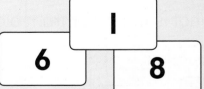

_____

_____

_____

66 sixty-six LXVI   11 + 11 + 11 + 11 + 11 + 11

Working with Hundreds, Tens, and Ones

Chapter 3 Lesson 8

NCTM Standards 1, 2, 6, 8, 9, 10

What number is shown by the blocks?

1.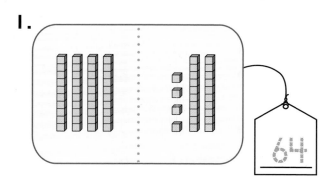

2.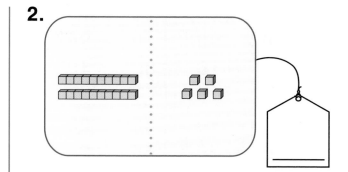

3.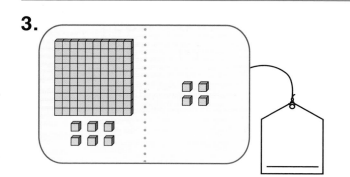

4.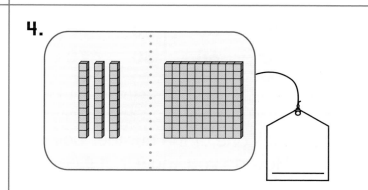

5.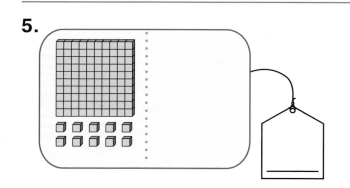

6.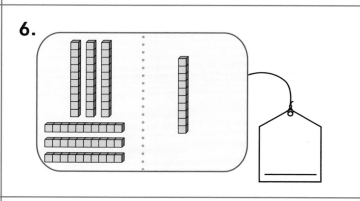

7.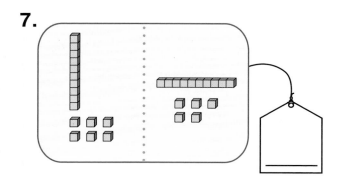

8.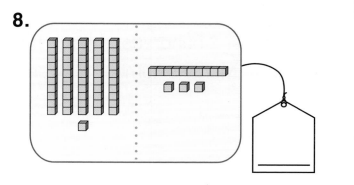

NOTE: Your child is learning that you name a number by combining hundreds with hundreds, tens with tens, and ones with ones.

70 − 3   **LXVII**   sixty-seven   **67**

9. Which pairs make 100? Circle them as fast as you can.

Sums of 100 Search

40/60	60/40	60/50	30/60	30/70	80/20	10/90	50/60	20/80
80/20	90/10	10/80	90/10	60/30	20/90	20/80	40/60	50/50
80/20	90/20	30/60	20/90	70/30	70/20	60/30	70/30	20/80
40/60	50/60	50/50	30/70	40/60	50/50	70/30	80/20	30/80

10. Pick one of the number pairs that you circled. How do you know that it has a sum of 100?

_____

_____

_____

Problem Solving

11. Marcy is thinking of a number. It has 6 tens, 4 ones, 2 hundreds. What is her number? _____

    Show Marcy's number in three different ways.

68 sixty-eight   LXVIII   34 + 34

Chapter 3 Lesson 9

Problem Solving Strategy
Draw a Picture
NCTM Standards 1, 2, 6, 7, 8, 9

- Understand
- Plan
- Solve
- Check

1. Mr. Brown's art room has 1 box of pencils, 2 packages of pencils, and 3 loose pencils. Ms. Gold's art room has 1 box and 5 packages of pencils. How many pencils are there in all?

   | 1 package = 10 pencils |
   | 1 box = 100 pencils |

   _____ pencils

2. There are 4 children in line. Laura is last. Charles is between Kim and Dan. Kim is next to Laura. Who is first in line?

   _____ is first.

3. Dogs, people, and fish live in a house. There are 8 heads and 18 legs. There are 3 dogs. How many people and fish are in the house? (Hint: A fish has no legs.)

   __3__ dogs _____ people _____ fish

NOTE: Your child is exploring different ways to solve problems. Sometimes drawing a picture is an efficient way to solve a problem.

Problem Solving Test Prep

1. The drawing shows how many pennies Tara saved each day. If the pattern continues, how many pennies will she save on Thursday?

Ⓐ 8   Ⓒ 12
Ⓑ 10   Ⓓ 16

2. Denise has 5 pages of 10 stickers and 3 loose stickers. Tom has 20 more stickers than Denise. How many stickers does Tom have?

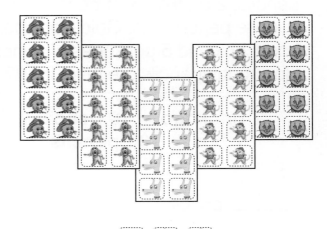

Ⓐ 35   Ⓒ 59
Ⓑ 37   Ⓓ 73

 Show What You Know

3. Ty asked 9 friends to name their favorite sport. Most like soccer best. The rest like football best. How many friends do you think like each sport?

_____ like soccer best.

_____ like football best.

Tell how you found your answer.

_____

4. Rob has 8 crayons. Lena has 7 more crayons than Rob. What number sentence can you use to find how many crayons Lena has?

_____

Tell how you found your answer.

_____
_____
_____

Chapter 3 Review/Assessment

NCTM Standards 1, 6, 7, 8, 9, 10

1. How many are there? Use the set of 10 to estimate. Then count to find the total. Lesson 1

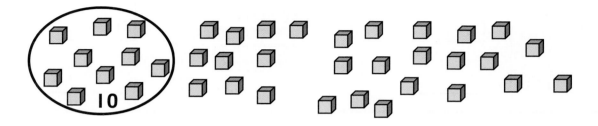

Estimate _____   Count _____

What number do the blocks show? Lesson 2

2.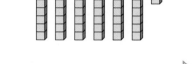

_____   _____ ⇒ _____
tens   ones

3.

_____   _____   _____ ⇒ _____
hundred tens   ones

What is missing? Draw symbols and write numbers. Lessons 3, 4

4.

7 tens 5 ones = _____ + _____ = _____

5.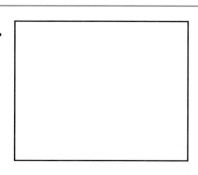

_____ hundreds _____ tens _____ ones

_____ + _____ + _____ = 309

70 + 1   **LXXI**   seventy-one   **71**

Show each number with the fewest blocks Draw symbols for the blocks. Lesson 5

6. 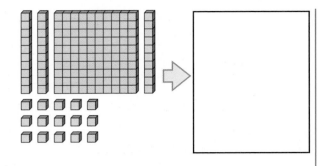 _____

7. 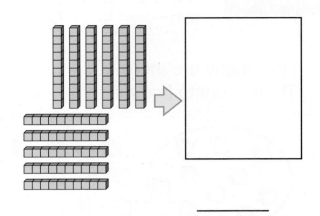 _____

Write the numbers. Then write <, >, or =. Lessons 6, 8

8.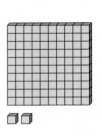

9.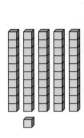

10. Write the numbers in order from biggest to smallest. Lessons 6, 7

150, 51, 75, 501, 455, 40

_____, _____, _____, _____, _____, _____

Problem Solving Lesson 9

11. Fran, George, Helena, and Jaime are in line for ice cream. Fran is third in line. George is between Jaime and Fran. Who is last in line?

_____

Name _____

Chapter 4: Addition and Subtraction with Place Value

Combining Blocks

You need
- 9 rods and 18 units

Put all of your base-ten blocks in a pile.

STEP 1 Making Sets

Choose any 6 base-ten blocks from the pile.
What blocks did you and your partner choose?

	How many rods?	How many units?	Total Value
Child 1			
Child 2			

STEP 2 Combining Sets

What is the value of the blocks in both sets? _____

Explain how you found the total.

_____

_____

STEP 3 Recording Totals

How can you record the way you combined the two sets of blocks?

_____

_____

_____

School-Home Connection

Dear Family,
   Today we started Chapter 4 of *Think Math!* In this chapter, I will work with base-ten blocks, symbols, and numbers to learn how to add and subtract. I will also work with dollars, dimes, and pennies. There are NOTES on the Lesson Activity Book pages to explain what I am learning every day.
   Here are some activities for us to do together at home. These activities will help me practice adding and subtracting.

Love,

_____

Family Fun

Adding Dimes and Pennies

Work with your child to practice addition using dimes and pennies.

- Gather a bunch of dimes and pennies for this activity.
- Each of you takes a handful of pennies and dimes and puts them in front of you.
- Work together to count each handful of coins and write the amounts on a sheet of paper. 43¢ 38¢
- Combine the coins from both handfuls.
- Work together to find how much money there is in all. If you have 10 pennies, trade them for a dime. Try counting the dimes first and then adding the pennies. Write the total.

  43¢
  38¢
  81¢

- Repeat with other handfuls of coins.

Cross Number Puzzles

Work with your child to complete a Cross Number Puzzle.

- A Cross Number Puzzle is like a crossword puzzle for numbers.
- Together, complete this puzzle by adding across and down. For each row, write the sum of the two numbers in the shaded box. Do the same for each column. In the green shaded box, write the sum of the rows and columns. The sum of the columns should always be the same as the sum of the rows!

4	8	
6	2	

4	8	12
6	2	8
10	10	20

74 seventy-four

Chapter 4 Lesson 1: Exploring Addition with Base-Ten Blocks

NCTM Standards 1, 2, 6, 7, 8, 9, 10

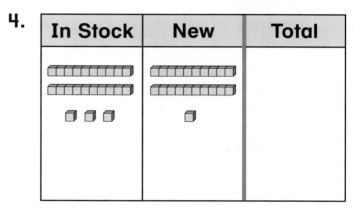

□ is 100, | is 10, and • is 1.

What is the total? Draw the symbols.

1. | In Stock | New | Total |
   |---|---|---|

2. | In Stock | New | Total |
   |---|---|---|

3. | In Stock | New | Total |
   |---|---|---|

4. | In Stock | New | Total |
   |---|---|---|

5. | In Stock | New | Total |
   |---|---|---|

6. | In Stock | New | Total |
   |---|---|---|

NOTE: Your child is learning to add hundreds, tens, and ones to find the sum. Your child can also record the blocks with symbols.

15    LXXV seventy-five 75

What is the total? Draw the symbols.

7. | In Stock | New | Total |
|---|---|---|
| | | |

8. | In Stock | New | Total |
|---|---|---|
| | | |

9. | In Stock | New | Total |
|---|---|---|
| | | |

10. | In Stock | New | Total |
|---|---|---|
| | | |

Problem Solving

11. Jack has 3 rods and 8 units. He wants to trade for a flat. What blocks does he need in order to make the trade? Use words, numbers, or pictures to explain.

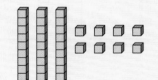

_____

_____

_____

_____

Chapter 4 Lesson 2: Exploring Subtraction with Base-Ten Blocks

NCTM Standards 1, 2, 6, 8, 9, 10

What is missing? Draw the symbols.

1. | In Stock | Sold | Left |
|---|---|---|
| (1 ten and 9 ones shown) | (5 ones) | (1 ten and 2 ones) |

2. | In Stock | Sold | Left |
|---|---|---|
| | (1 ten) | (4 ones) |

3. | In Stock | Sold | Left |
|---|---|---|
| | (1 ten and 3 ones) | (1 ten and 1 one) |

4. | In Stock | Sold | Left |
|---|---|---|
| (2 tens and 6 ones, plus 1 ten on top) | (1 ten and 2 ones) | |

5. | In Stock | Sold | Left |
|---|---|---|
| (2 tens on top, 2 tens and 8 ones) | (1 ten and 6 ones) | |

6. | In Stock | Sold | Left |
|---|---|---|
| (tens arranged) | | (3 tens) |

NOTE: Your child is learning to subtract with hundreds, tens, and ones. Ask your child to explain how to find the answer for Problem 6.

7 × 11   LXXVII   seventy-seven   77

What is missing? Draw the symbols.

7.
In Stock	Sold	Left
1 ten 3 ones	4 ones	9 ones

"Open a package if you need ones."

8.
In Stock	Sold	Left
1 ten		4 ones

9.
In Stock	Sold	Left
1 ten 3 ones	3 ones	

10.
In Stock	Sold	Left
1 ten 5 ones	7 ones	

11.
In Stock	Sold	Left
2 tens	6 ones	

Challenge

12. What is missing?

In Stock	Sold	Left
1 hundred	3 ones	

78 seventy-eight LXXVIII △ 39 + 39

Chapter 4 Lesson 3

Wonder Wheel Addition and Subtraction
NCTM Standards 1, 2, 6, 8, 9, 10

What is the total? Draw the symbols.

1.

In Stock	New	Total

2.

	In Stock
	New
	Total

3.

	In Stock
	New
	Total

4.

	In Stock
	New
	Total

NOTE: Your child is learning to add and subtract two- and three-digit numbers. Your child can also record the blocks with symbols.

What is missing? Draw the symbols.

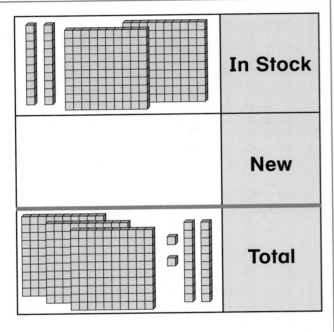

Challenge

9. Show two different ways to make the same total. Draw the symbols.

80 eighty   LXXX   80   16

Name _____ Date _____

Chapter 4
Lesson 4

Introducing the Cross Number Puzzle

NCTM Standards 1, 2, 6, 7, 8, 9, 10

What is missing?

Rules:
- Make the same amount on both sides of the thick line.
- You cannot have more than 9 of any symbol in each box.

Work across and down. Follow the rules.

1.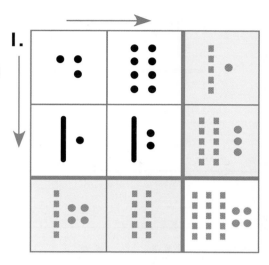

2.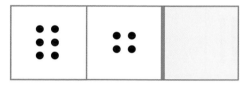

3.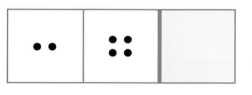

4.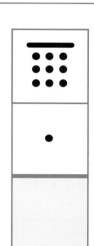

5.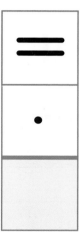

6.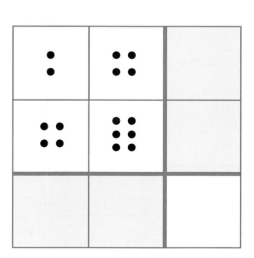

7.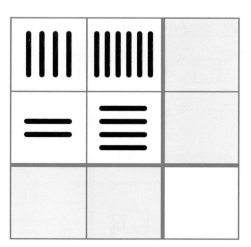

NOTE: Your child is learning to solve Cross Number Puzzles using symbols. These puzzles will help your child understand how to add and subtract numbers.

90 − 9 ☐ 81   LXXXI   eighty-one   81

What is missing?

8.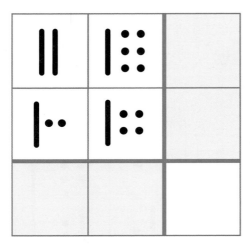

9.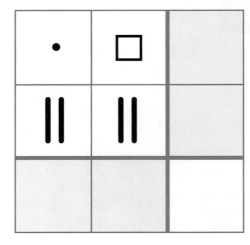

10.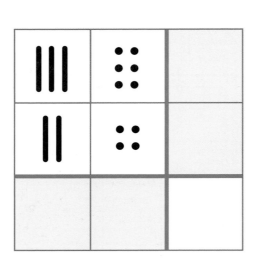

11. Make your own.
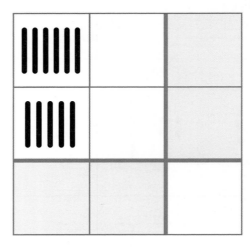

Challenge

12. What is missing?   How can you check your answers?

_____
_____
_____
_____
_____

Addition and Subtraction

Chapter 4 Lesson 5

NCTM Standards 1, 2, 6, 9, 10

What is missing? Draw the symbols and write the numbers.

1.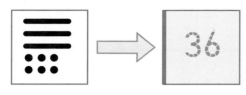

2.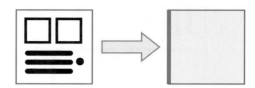

3.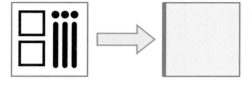

4.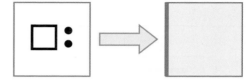

5.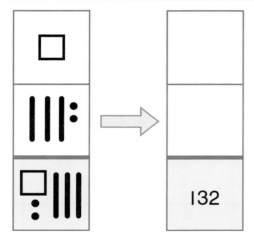

6.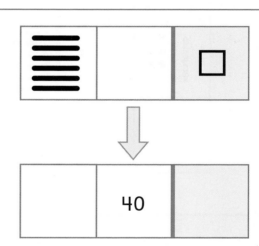

7.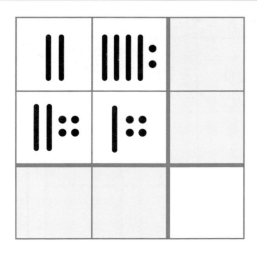

NOTE: Your child is learning to add and subtract two- and three-digit numbers using Cross Number Puzzles.

80 + 3   **83**   LXXXIII   eighty-three   83

What is missing? Draw the symbols and write the numbers.

8.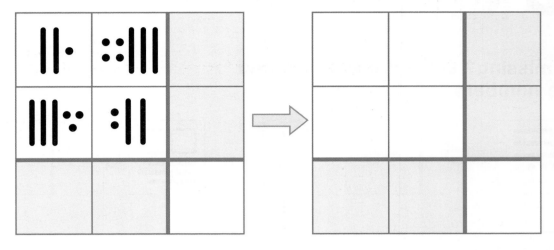

9.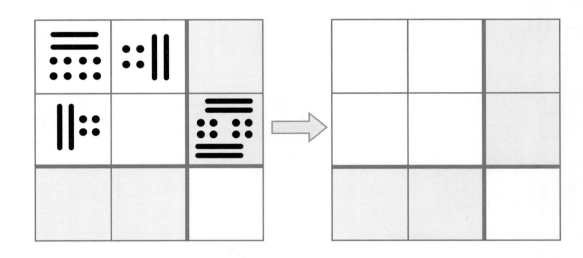

Challenge

10. What is missing?

	☐	

→

20	4	
	9	

84 eighty-four   LXXXIV   7 dozen

Adding a Multiple of 10

Chapter 4 Lesson 6

NCTM Standards 1, 2, 6, 7, 8, 9, 10

What is missing?

1.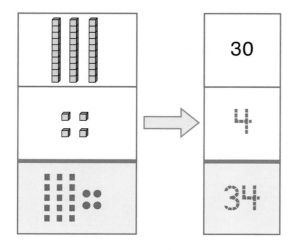

2.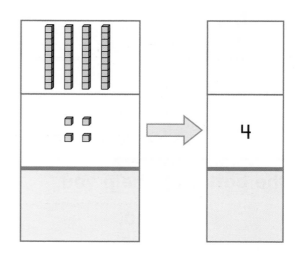

3.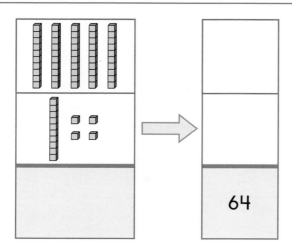

4.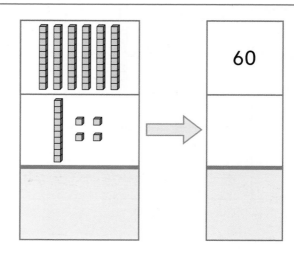

5.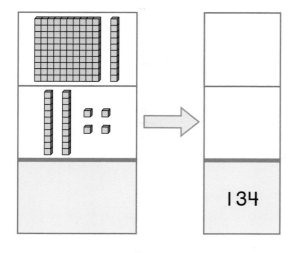

6.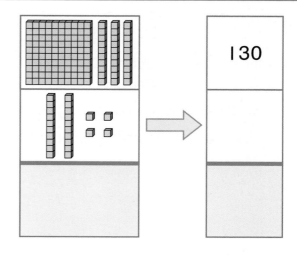

NOTE: Your child is learning to add multiples of 10. Ask your child to solve 20 + 16.

What is missing?

7.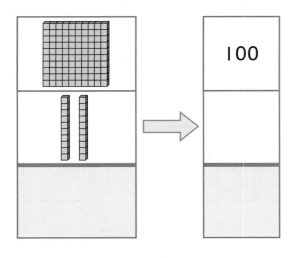

8.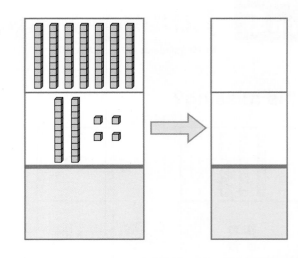

Use the pattern to help you.

9.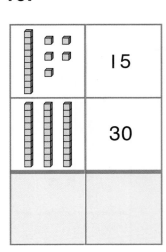

10.

11.

12.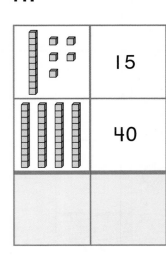

Problem Solving

13. Mrs. Sung buys a roll of 100 stamps. She has 32 stamps at home. How many stamps does Mrs. Sung have in all?

    _____ stamps

    Use words, numbers, or pictures to explain.

Chapter 4 Lesson 7: Fewest Dimes and Pennies

NCTM Standards 1, 2, 6, 7, 8, 9, 10

How can you show 24¢ using only dimes and pennies?

24¢	
Dimes	Pennies
0	24
1	14
2	4

This way uses the fewest coins.

How can you show each money amount with the fewest coins? Use only dimes and pennies.

Money Amount	How many dimes?	How many pennies?	How much money?
1. twenty-four cents	2	4	24 ¢
2. twenty-five cents			¢
3. twenty-six cents			¢
4. thirty-six cents			¢
5. forty-six cents			¢
6. fifty-two cents			¢
7. seventeen cents			¢
8. ten cents			¢

NOTE: Your child is learning to represent money amounts using the fewest dimes and pennies. Together, show 43¢ using the fewest dimes and pennies.

90 – 3   LXXXVII   eighty-seven   87

Show each money amount as many ways as you can.
Use only dimes and pennies.

9.
16¢	
Dimes	Pennies
0	16

How many coins are used to show 16¢ with the fewest dimes and pennies?

_____ coins

10.
37¢	
Dimes	Pennies

How many coins are used to show 37¢ with the fewest dimes and pennies?

_____ coins

11. Max has 42¢ in dimes and pennies. He has the fewest coins for that amount. How many coins does he have? Tell how you know.

_____ coins

_____

_____

Problem Solving

12. Tara has only pennies and dimes in her pocket. She takes out 5 coins. What is the most money that Tara could have? Explain.

_____ ¢

_____

Chapter 4 Lesson 8

Fewest Dollars, Dimes, and Pennies

NCTM Standards 1, 2, 6, 7, 8, 9, 10

How can you show $2.43 with the fewest bills, dimes, and pennies?

Use as many dollar bills as you can. Use as many dimes as you can.

2 dollars + 4 dimes + 3 pennies = $2.43

How can you show each money amount with the fewest bills and coins? Use only dollars, dimes, and pennies.

	Money Amount	How many dollars?	How many dimes?	How many pennies?
1.	$1.62	1	6	2
2.	$1.57			
3.	$1.93			
4.	$2.14			
5.	$2.30			
6.	$3.41			
7.	$3.08			

NOTE: Your child is learning to represent money amounts with the fewest dollars, dimes, and pennies. Together, show $1.23 using 1 bill, 2 dimes, and 3 pennies.

90 − 1 LXXXIX eighty-nine 89

How much money is there?

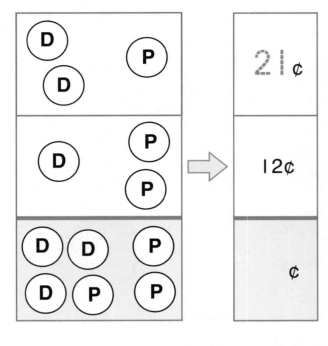

9. [Image shows: top box with 1 D and 5 P = ___¢; middle box with 1 D and 5 P = ___¢; bottom box with 3 D = ___¢]

10. Joanne has $2.63. What is the fewest bills and coins she could have? Explain how you know.

_____

_____

_____

Problem Solving

11. Jared has 1 dollar bill, 4 dimes, and 16 pennies. How much money does he have?

$ _____

Can he show the amount with fewer dollars, dimes, and pennies? Use words, numbers, or pictures to explain.

Chapter 4 Lesson 9

Problem Solving Strategy
Work Backward
NCTM Standards 1, 2, 6, 7, 8, 9, 10

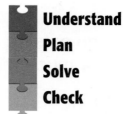

Understand
Plan
Solve
Check

1. Brad's basketball team scored 14 points in the second half. They had 56 points at the end of the game. How many points did the team get in the first half?

   _____ points

   Explain how you found your answer. _____

   _____

   _____

2. Jen reads every night for 20 minutes. Tonight she finishes at 8:00. What time did she start reading?

   ___ : ___

   Explain how you found your answer. _____

   _____

   _____

3. The bus stops and 8 people get on and 10 people get off. Now there are 19 people on the bus. How many people were on the bus before it stopped?

   _____ people

   Explain how you found your answer. _____

   _____

   _____

NOTE: Your child is exploring different ways to solve problems. Sometimes *working backward* is an efficient way to solve a problem

90 + 1   XCI   ninety-one 91

Problem Solving Test Prep

1. Juan puts 5¢ a day in his bank. How much will he save in 6 days?

   Ⓐ 5¢

   Ⓑ 15¢

   Ⓒ 30¢

   Ⓓ 35¢

2. Mary took 3 pencils out of a new box. There are 15 pencils left in the box. How many pencils were in the box to start?

   Ⓐ 3 pencils

   Ⓑ 9 pencils

   Ⓒ 12 pencils

   Ⓓ 18 pencils

Show What You Know

3. Ann, Tim, and Carlos are in a race. How can they finish if a girl comes in first?

   _____   _____   _____
    first   second   third

   Explain your answer.

   _____

   _____

   _____

   _____

4. Lee has only pennies and dimes. He has 6 coins. What is the most money he can have?

   _____ ¢

   Explain your answer.

   _____

   _____

   _____

Chapter 4 Review/Assessment

NCTM Standards 1, 2, 6, 7, 8, 9, 10

What is missing? Draw the symbols. Lessons 1–4

1. | In Stock | New | Total |
|---|---|---|
| (2 tens, 8 ones) | (7 tens, 5 ones) | |

2. | In Stock | Sold | Left |
|---|---|---|
| (1 ten, 3 ones) | (8 ones) | |

3. | In Stock | New | Total |
|---|---|---|
| | (5 tens, 7 ones) | (1 hundred) |

4. | | | |
|---|---|---|
| (1 ten, 4 ones) | | In Stock |
| (1 ten, 6 ones) | | New |
| | | Total |

5. **What is missing? Draw the symbols and the numbers.** Lesson 5

||| :: ⁞		··	
|| ·	||| ·	

→

36		

3 × 31   **XCIII**   ninety-three   **93**

What is missing? Lesson 6

6.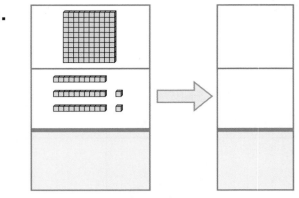

7.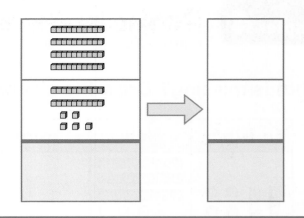

How can you show each amount with the fewest coins? Lesson 7

	Money Amount	How many dimes?	How many pennies?	How much money?
8.	thirty-six cents			¢
9.	ninety-seven cents			¢

How much money is there? Lesson 8

10.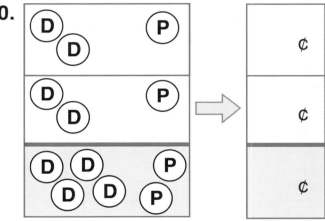

11.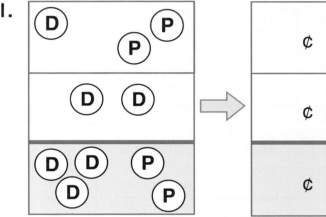

Problem Solving Lesson 9

12. Greg bought 12 toy cars. Now Greg has 28 cars. How many cars did Greg have before he bought these new ones? _____ cars

    Explain how you found your answer. _____

    _____

Chapter 5: Probability and Data
Probability of Pulling Pairs

You need
- two pairs of cubes in different colors
- paper lunch bag

What is the chance you will pull out two cubes that are the same color?

STEP 1 Pulling Out Two

Pull out two cubes. Are both cubes the same color?
Use a ✔ to record each turn.

Matching Pair	
Yes	No

Will you always pull out a matching pair? Explain. _____

STEP 2 Pulling Out Three

Pull out three cubes. Are two of the cubes the same color? Record each turn.

Matching Pair	
Yes	No

Tell about your results. _____

Why do you think this happens? _____

_____

STEP 3 Pulling Out Four

Pull out four cubes. Will you get a matching pair? Explain.

_____

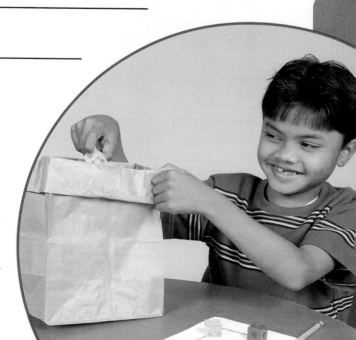

School-Home Connection

Dear Family,

Today we started Chapter 5 of *Think Math!* In this chapter, I will explore probability, picture graphs, real-object graphs, bar graphs, pictographs, and even line graphs. There are NOTES on the Lesson Activity Book pages to explain what I am learning every day.

Here are some activities for us to do together at home. These activities will help me understand probability and data.

Love,

_____

Family Fun

What Are You Pulling?

Work with your child to play a probability game called *What Are You Pulling?*

- Put 30 of the same food item into a paper lunch bag. You might use candies, cereal pieces, or crackers in different colors or shapes. Do not let your child see what is inside the bag.

- Take turns with your child. Pull out one item at a time from the bag without looking inside. Then replace the item and mix them up. Record the results of each turn. After taking 10 turns, start asking this question: What do you think you will pull out now?

- With each turn, your child should get better at making predictions. Look back at what has already been pulled out before making a prediction.

- Continue for as long as you wish. Then, if you want, eat the items together!

Word Search

Work with your child to circle these words in the puzzle.

BAR GRAPH   DATA
IMPOSSIBLE   KEY
PICTOGRAPH   LIKELY
PICTURE GRAPH   POSSIBLE
LINE GRAPH   PREDICT
SYMBOL   TALLY

P	E	S	Y	L	L	D	Y	I	M	T
I	B	A	R	G	R	A	P	H	L	A
C	M	O	E	S	H	T	O	E	I	L
T	A	L	L	Y	H	A	S	R	K	I
U	L	A	K	E	Y	P	S	A	E	N
R	I	M	P	O	S	S	I	B	L	E
E	N	A	P	H	G	E	B	B	Y	G
G	O	S	Y	M	B	O	L	A	N	R
R	A	P	H	A	G	R	E	S	I	A
A	P	R	E	D	I	C	T	O	T	P
P	I	C	T	O	G	R	A	P	H	H
H	M	T	U	K	L	I	T	R	U	N

Chapter 5 Lesson 1

Exploring Probability

NCTM Standards 1, 2, 5, 6, 7, 8, 9, 10

There are 10 cubes in each bag. Some are red and some are blue. Write how many of each and color the cubes to match the story.

1. Joe is more likely to pull out a red cube than a blue cube from the bag.

   _____ red   _____ blue

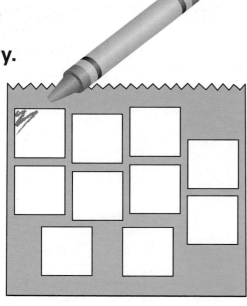

2. Shamari is equally likely to pull out a red cube or a blue cube.

   _____ red   _____ blue

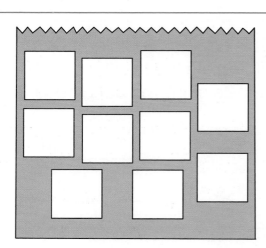

3. Make your own story.

   Matt is _____ likely

   to pull out a _____ cube

   than a _____ cube.

   _____ red   _____ blue

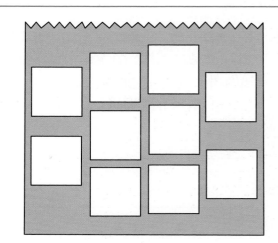

NOTE: Your child is learning about the likelihood of events. Talk about events that are more likely and less likely to happen in your everyday lives.

100 − 3   XCVII   ninety-seven   97

Each bag has different cubes. Sue pulls out a cube and records the color. She does this 20 times. She puts the cube back each time. Which color do you think she is more likely to pull out next?

4.

Red	Blue
⊦⊦⊦⊦ ⊦⊦⊦⊦ ⊦⊦⊦⊦ III	II

_____

5.

Red	Blue
⊦⊦⊦⊦	⊦⊦⊦⊦ ⊦⊦⊦⊦ ⊦⊦⊦⊦

_____

6.

Red	Blue
	⊦⊦⊦⊦ ⊦⊦⊦⊦ ⊦⊦⊦⊦ ⊦⊦⊦⊦

_____

7. This bag has 5 red and 5 blue cubes. Complete the tally table to show possible results of pulling out a cube 20 times.

Red	Blue

Problem Solving

8. Sue pulled out a cube from a bag 20 times. She made this tally table. What are the chances that a green cube was in this bag? Explain.

Red	Blue
⊦⊦⊦⊦ I	⊦⊦⊦⊦ ⊦⊦⊦⊦ IIII

_____

_____

_____

98 ninety-eight   XCVIII   49 + 49

Chapter 5
Lesson 2
Using Real-Object Graphs and Picture Graphs

NCTM Standards 1, 2, 4, 5, 6, 7, 8, 9, 10

Each child in Room 2 has a penny. The children lined up to show the years their pennies were minted. They recorded the data like this.

The mint year is the year a coin was made.

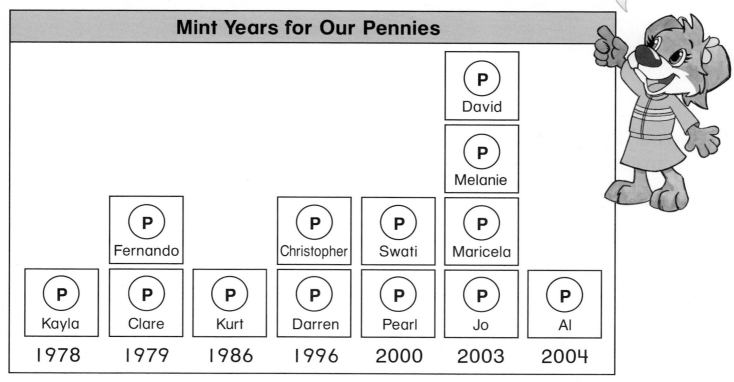

1. The oldest penny was minted in _____.

2. _____ has the newest penny.

3. There are _____ children in Room 2.

4. Melanie's penny was minted in _____.

 5. Write your own sentence about the picture graph.

 NOTE: Your child is learning to make and use real-object graphs and picture graphs to show information.

9 × 11 **XCIX** ninety-nine **99**

Hair Color of Children in Room 2					
Brown	Swati	Al	David	Fernando	
Blonde	Kayla	Christopher	Clare		
Black	Jo	Kurt	Darren	Maricela	Melanie
Red	Pearl				

✏️ **Write two questions that can be answered from the graph. Give the answer.**

6. _____

7. _____

✏️ **8.** Write a question that cannot be answered from the graph.

_____

Challenge

9. I am _____ years old. I was born in _____.

In 2020, I will be _____ years old.

100 one hundred

Chapter 5 Lesson 3

Using Bar Graphs to Investigate Probability

NCTM Standards 1, 2, 4, 5, 6, 7, 8, 9, 10

You have two number cubes. One cube has numbers 1 to 6. The other has numbers 7 to 12.

1. What are all of the possible addition facts you could make by tossing the number cubes?

+	7	8	9	10	11	12
1	1 + 7					
2		2 + 8				
3			3 + 9			
4						
5						
6						

2. What are all of the possible sums?

_____

3. How many different ways can you make each sum by tossing the number cubes? Tally to record.

Sum	8	9	10	11	12	13	14	15	16	17	18
Tally	I	II									
Number of Ways	1	2									

 NOTE: Your child is learning about bar graphs and is using this representation to record the results of an addition and probability game.

Play the game with a classmate. Toss two number cubes. Color a box in the graph for each sum. Which numbers win? Answer Questions 4 and 5 as you play. Answer Question 6 at the end.

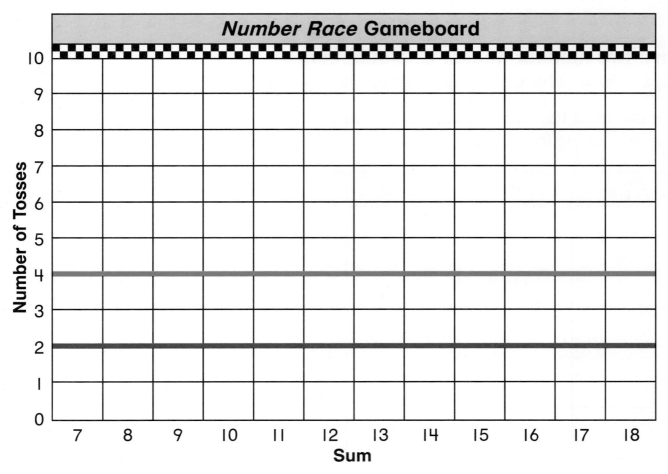

4. Which number got to the blue line first? _____

5. Which number got to the green line first? _____

6. Which number finished the race first? _____

Challenge

7. How could you change the game so that one sum is certain to win every time?

_____
_____
_____

Chapter 5 Lesson 4: Making and Using Bar Graphs

NCTM Standards 1, 2, 4, 5, 6, 7, 8, 9, 10

This class list gives the names for all children in Room 2.

✓ Melanie	Christopher	Kayla	Clare
Jo	Kurt	Swati	Maricela
Fernando	Darren	Pearl	Al
David			

Check off each name as you record it in the graph.

1. Use the class list. Make a bar graph showing the lengths of the names.

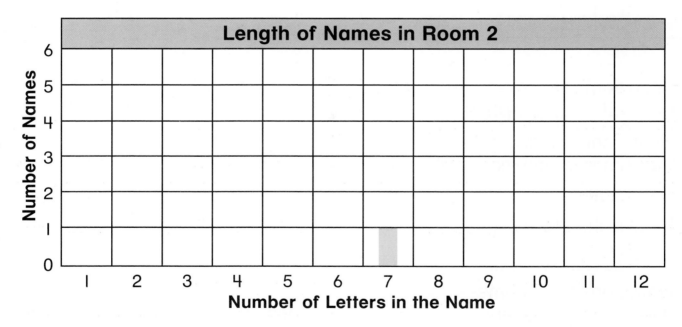

2. How many letters are in the longest name? _____ letters

3. How many names have 5 letters? _____ names

NOTE: Your child is learning how to create a bar graph. Your child is also learning to identify whether a question can be answered by a table or a graph.

100 + 3   103   CIII   one hundred three   103

4. Use the table to make a bar graph. Remember to label the graph.

Length of Names in Room 3	
Number of Letters	Name
3	Ned
4	Anna John Ling
5	Kayla Nikil
7	Chelsea Tabitha
8	Benjamin
9	Elizabeth

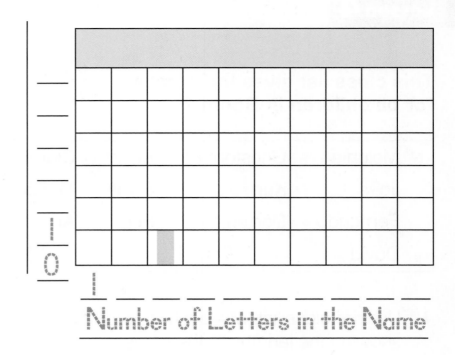

Number of Letters in the Name

Answer each question. Circle whether you use the table or graph to find the answer.

5. How many letters are in the names for the most children? _____ letters   table   (graph)

6. What is the shortest name? _____   table   graph

7. How many more names have 4 letters than 9 letters? _____ names   table   graph

Challenge

8. What can you learn from the table that is not in the graph?

_____

_____

Name _____   Date _____

Chapter 5
Lesson 5

Making and Using Pictographs
NCTM Standards 1, 2, 5, 6, 7, 8, 9, 10

Tom asked the children in each grade at school whether they have a dog. Then he made this pictograph.

Children at Bayles School with Dogs						
			☺		
			☺		
		☺	☺		☺
☺		☺	☺		☺
☺		☺	☺		☺
☺	☺	☺	☺	☺	☺
1st	2nd	3rd	4th	5th	6th

Grade

Key: Each ☺ stands for 5 children.

1. How many first graders have dogs? _____ children

   How does the graph show that? _____

   _____

2. How many more fourth graders than third graders have dogs? _____ children

   How did you figure that out? _____

   _____

NOTE: Your child is learning how to create a pictograph where a symbol stands for more than one piece of data.

3. Use the tally table to make a pictograph. Remember to choose a symbol and make a key.

Children at Bayles with Cats	
Grade	Tally
1													
2														
3								
4			
5																
6						

Children at Bayles School with Cats

Grade								
1st								
2nd								
3rd								
4th								
5th								
6th								

Key: Each _____ stands for _____ children.

4. Write your own sentence about the pictograph.

_____

_____

Write two questions that can be answered from the graph.

5. _____

6. _____

Problem Solving

7. Three children in Kindergarten have cats. How could you show this in the graph if each symbol stands for 2 children?

_____

_____

106 one hundred six   CVI   53 + 53

Chapter 5
Lesson 6: Graphing Change Over Time

NCTM Standards 1, 2, 4, 5, 6, 7, 8, 9, 10

This table shows typical snowfall for Chicago, Illinois. The amount of snow for March has been left out.

Snowfall in Chicago							
Month	Oct	Nov	Dec	Jan	Feb	Mar	Apr
Inches of Snow	$\frac{1}{2}$	2	8	11	8	?	2

1. Make a line graph of the data in the table.

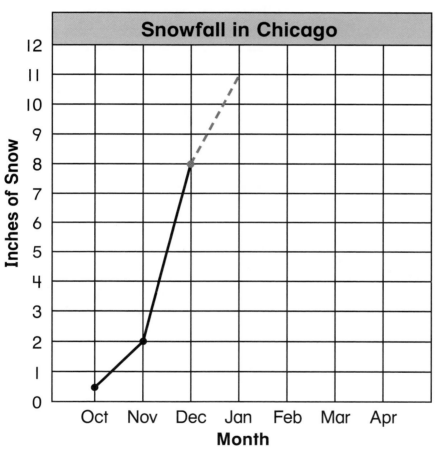

Find the point on the graph where the month and the inches intersect.

2. Use the graph to estimate how much snow fell in March. Explain your answer.

_____

_____

NOTE: Your child is learning how to create and interpret line graphs to show change over time.

110 − 3   **CVII**   one hundred seven   107

Draw a line from each story to the graph that matches it best.

3. Lakisha saved money in her piggy bank. She sometimes spent some of it.

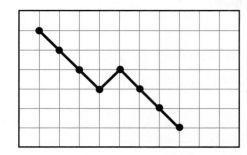

4. The teacher in Room 2 recorded how many children came to school each day.

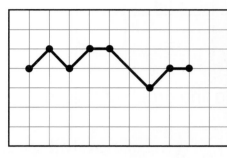

5. Al completed the same addition table several times. He recorded how long it took each time.

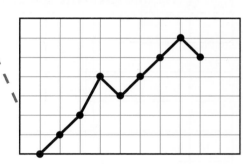

6. How are a line graph and a bar graph different?

_____

_____

Problem Solving

7. Write a story about this line graph. Think about what makes sense for a line that keeps going up.

_____

_____

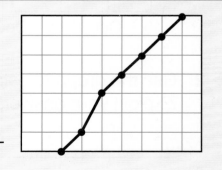

Chapter 5 Lesson 7

Problem Solving Strategy
Make a Table

NCTM Standards 1, 2, 4, 5, 6, 7, 8, 9, 10

Understand · Plan · Solve · Check

1. Nev spins both spinners shown. What sum is she most likely to get?

   Spinner A   Spinner B

   _____

Spinner A	1	1	1	3	3	3	4	4	4
Spinner B	4	5	7	4	5	7			
Sum	5								

2. How many tires will Seth need for 7 model cars?

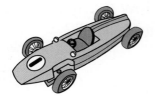

Cars	1	2	3				
Tires							

_____ tires

3. Cindy has 3 coins in her pocket. She only has coins worth 5¢ or less. How much money might she have?

Pennies	Nickels	Total Money
0	3	

NOTE: Your child is exploring different ways to solve problems. Sometimes making a table is an efficient way to solve a problem.

110 − 1   CIX   one hundred nine 109

Problem Solving Test Prep

1. Kip has 12 blocks and 2 plates. He puts half of the blocks on each plate. How many blocks are on each plate?

    Ⓐ 24 blocks

    Ⓑ 12 blocks

    Ⓒ 6 blocks

    Ⓓ 3 blocks

2. Sara had 25¢ in her bank. She put some more coins into the bank. Then she had 38¢. How much money did she put in the bank?

    Ⓐ 63¢

    Ⓑ 13¢

    Ⓒ 5¢

    Ⓓ 3¢

Show What You Know

3. Ivana has 32 marbles in a solid color and 15 that are striped. How many marbles does she have in all?

    _____ marbles

    Explain how you found the answer.

    _____

    _____

    _____

    _____

4. Carlos drove 183 miles on Saturday. He drove farther on Sunday. The number of miles has the same digits in a different order. How far did Carlos drive on Sunday?

    _____ miles

    Explain how you found the answer.

    _____

    _____

    _____

    _____

Name _____ Date _____

Chapter 5 Review/Assessment
NCTM Standards 1, 2, 4, 5, 6, 7, 8, 9, 10

1. Which color is more likely to be pulled out next from the bag? Use the tally table to help. **Lesson 1**

Red	Blue

_____

Use the picture graph to complete each sentence. Lesson 2

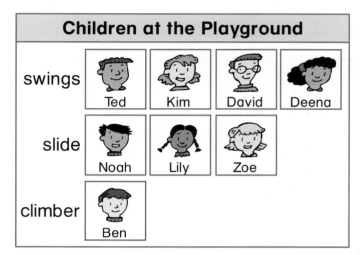

2. There are _____ children at the playground.

3. _____ is on the climber.

4. There are _____ more children on the slide than on the climber.

Use the bar graph to answer each question. Lessons 3, 4

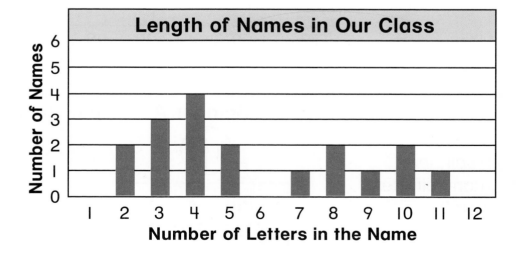

5. How many letters are in the longest name? _____ letters

6. How many names are 4 letters long? _____ names

3 × 37   CXI   one hundred eleven 111

7. Use the tally table to make a pictograph. Choose a symbol and make a key. Lesson 5

Crayons in the Box	
Color	Tally
blue	卌 卌 ∥
green	卌 卌 卌 ∣
red	卌 卌 ∥
yellow	卌 ∥∥

Crayons in the Box								
blue								
green								
red								
yellow								

Key: Each _____ stands for _____ crayons.

Answer each question about the line graph. Lesson 6

8. How many hot dogs were sold at 3:00?

_____ hot dogs

9. Did hot dog sales go up or down from 4:00 to 6:00?

_____

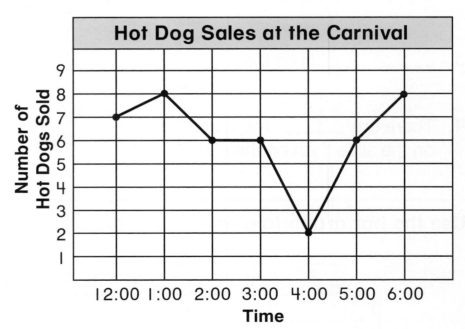

Problem Solving Lesson 7

10. Sal spins both spinners shown. How many different sums can he spin?

_____ sums

Spinner A   Spinner B

Spinner A	1	1	3	3
Spinner B	1	2	1	2
Sum				

112   one hundred twelve   CXII   56 + 56

Name _____

Chapter 6

Measuring Time
School Schedule

You need
- index cards
- crayons or markers

Discuss the activities you do in school.

STEP 1 **Describing Activities**

What activities happen every day in school?

_____

_____

_____

STEP 2 **Ordering Activities**

Draw each activity on a card. In what order do you do the activities in school? Put the cards in order starting with the first activity of the day.

STEP 3 **Thinking About Time**

Which activities seem to take a long time?

_____

_____

Which activities seem to take a short time?

_____

_____

Investigation

School-Home Connection

Dear Family,

Today we started Chapter 6 of *Think Math!* In this chapter, I will learn to read time. I will also solve problems involving time. There are NOTES on the Lesson Activity Book pages to explain what I am learning every day.

Here are some activities for us to do together at home. These activities will help me understand concepts related to time.

Love,

_____

Family Fun

How Much in 5 Minutes?

Have your child complete activities that take 5 minutes.

- Ask your child, "What do you think you could do in 5 minutes?" Make a list. Possibilities include: make my bed; write the alphabet; and practice my math facts.

- Ask your child to choose one of the activities on the list to do as you time it. How long did it take?

- Repeat the activity for several days in a row. Together, graph the amount of time the activity takes every day.

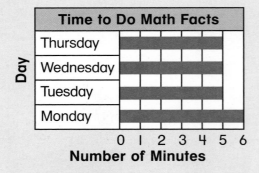

What Time Is It?

Work with your child to practice telling time.

- Discuss with your child the activities your family does every day. For example, get up in the morning, get washed and dressed, eat meals, come home, go to bed, and so on.

- Whenever a new activity begins or ends, ask your child, "What time is it?" Have your child read a clock or a watch to tell the time. Repeat this activity throughout the day.

It's 7:00. Time to wake up!

Name _____ Date/Time _____

Chapter 6 Lesson 1

Exploring Time

NCTM Standards 4, 5, 6, 7, 8, 9, 10

1. How many smiley faces can you draw in 10 seconds? Have a partner time 10 seconds as you draw.

How many smiley faces did you draw? _____ faces

2. How many smiley faces can you draw in 1 minute? Have a partner time 1 minute as you draw.

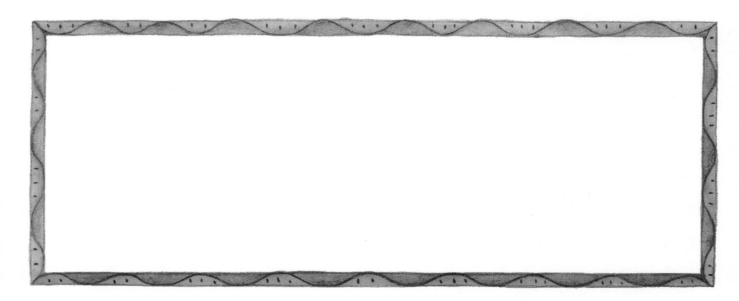

How many smiley faces did you draw? _____ faces

 NOTE: Your child is learning different units of time including seconds, minutes, and hours.

 CXV one hundred fifteen 115

3. Draw a picture of something you do that takes about 1 hour.

4. Does an hour seem long or short to you? Tell why.

_____

_____

_____

Challenge
Circle the better estimate for each time.

5. About how long does it take to say the alphabet?

    1 second   1 minute

6. About how long does it take to play a game of soccer?

    1 minute   1 hour

Chapter 6 Lesson 2

Minutes in an Hour

NCTM Standards 4, 6, 8, 10

Name _____   Date/Time _____

What time is it?

1.    5:00

2.    :

3.    :

4.    :

5. 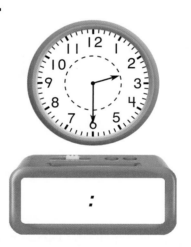   :

6.    :

7. Show any time to the half hour. Draw the clock hands and write the time.

NOTE: Your child is learning to tell time to the hour and half hour. Ask your child to read the clock when the time is on the hour.

120 − 3   **CXVII**   one hundred seventeen   117

What is missing?

8.  1 hour later

Draw the clock hands.

3:30

:

9.   2 hours later

:   :

Problem Solving

10. A machine makes 1 toy every minute. How many toys can the machine make in 25 minutes? _____ toys

11. A machine makes 2 boxes every minute. How many boxes can the machine make in 1 hour? Explain your answer. _____ boxes

_____

_____

_____

118 one hundred eighteen   CXVIII   59 + 59

Chapter 6
Lesson 3

Telling Time to 10 Minutes

NCTM Standards 3, 4, 6, 7, 8, 9, 10

What time is it?

1.

3 :

2.

 :

3.

 :

4.

20 minutes after 8

5.

_____ minutes after _____

6.

_____ minutes before _____

7.

_____ minutes before _____

NOTE: Your child is learning to tell time to 10 minutes and describe the time as before and after the hour.

120 − 1   **CXIX**   one hundred nineteen   **119**

8. Tina travels 1 mile every 10 minutes. What time will she get to each mile flag? Draw the clock hands.

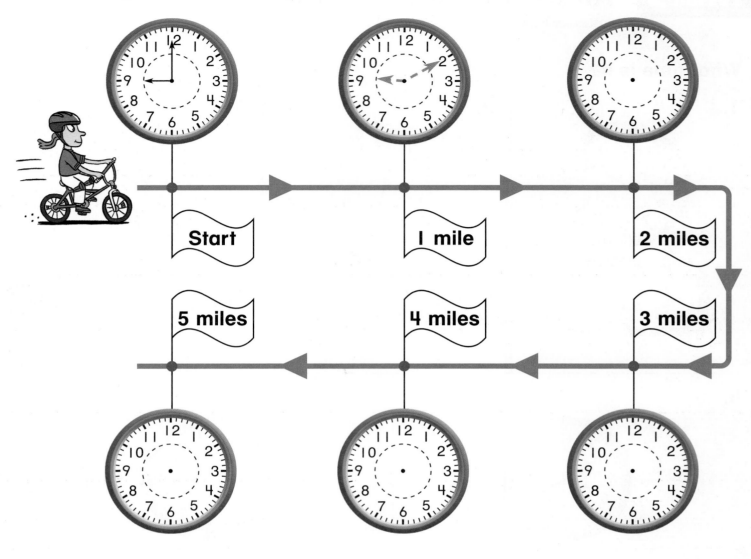

Problem Solving

9. Dana starts her homework at 3:10. It takes 20 minutes to finish. What time is she done? ____:____

   Use words, numbers, or pictures to explain.

Chapter 6 Lesson 4: How Far? How Fast?

NCTM Standards 3, 4, 6, 8, 9, 10

What is missing?

1.   30 minutes later

2.   half an hour later

3.  1 hour later

4.  1 and a half hours later

5. Make your own.

    2 and a half hours later

NOTE: Your child is learning to read and write time. Ask your child what time it will be in 1 hour.

6. Nick walks 2 miles in 1 hour. He starts at 4:00. What time will he get to each mile flag?

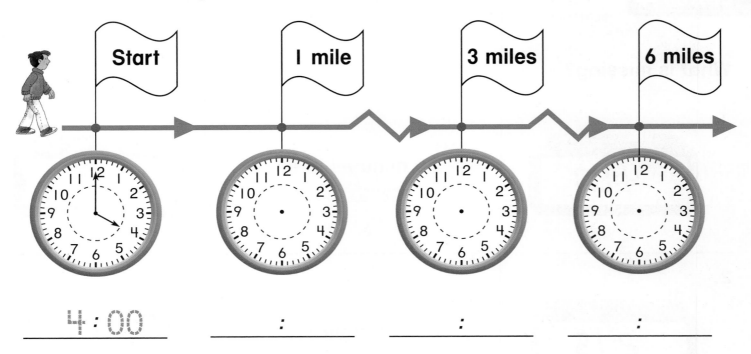

4:00   :   :   :

7. How can you write 8:15 in words? Write it in two different ways.

_____

_____

Problem Solving

8. A movie starts at 6:50. It is 2 and a half hours long. What time will the movie end? _____ :

Use words, numbers, or pictures to explain.

122 one hundred twenty-two   CXXII   61 + 61

**Chapter 6
Lesson 5**

Telling Time to 5 Minutes

NCTM Standards 4, 8, 9, 10

Exactly what time is it?

1.

2.

3.

4.

5.

6.

About what time is it? Circle the closer time.

7.

   5:00   5:30

8.

   1:30   2:00

9.

   7:30   8:00

 NOTE: Your child is learning to tell time to 5 minutes and to estimate time. At a time right after the hour, ask your child, "About what time is it?"

100 + 20 + 3   **CXXIII**   one hundred twenty-three   123

What is missing?

10.

 5 minutes later →

11.

 15 minutes later →

Challenge

12. What is missing?

5 minutes earlier			2:15	4:05	
Time	9:30	11:15	2:20		
15 minutes later		11:30			5:15

Chapter 6
Lesson 6

Telling Time to the Minute

NCTM Standards 1, 4, 6, 8, 9, 10

What time is it?

1.

2.

Draw the clock hands.

3.

4.

5.

6.

NOTE: Your child is learning to tell time to the minute. Ask your child to tell the exact time several times during the day.

25 **CXXV** one hundred twenty-five 125

What landmark on the clock can help you find the time? Write the landmark and the exact time.

7.

landmark time __8:30__

exact time __8:28__

8.

landmark time _____:_____

exact time _____:_____

9.

landmark time _____:_____

exact time _____:_____

10.

landmark time _____:_____

exact time _____:_____

11. In Problem 9, how did you use the landmark to find the exact time? _____

_____

_____

Challenge

12. What time is it? How do you know? _____:_____

_____

_____

Chapter 6
Lesson 7

Calendar and Ordinal Numbers
NCTM Standards 1, 4, 6, 8, 9, 10

1. Write the name of the month and the year.
   Write the dates in the calender.

   _____ , _____

Sunday	Monday	Tuesday	Wednesday	Thursday	Friday	Saturday

2. How many days are in this month? _____ days

3. What day is the first of the month? _____

4. What day is the nineteenth of the month? _____

5. What is the date of the third Monday? _____

NOTE: Your child is learning to read a calendar and work with ordinal numbers. Together, talk about the dates of holidays and events this month.

130 − 3   **CXXVII**   one hundred twenty-seven   127

What floor is it?

6. Draw an X on the box for the tenth floor.

The boxes on the side are elevator floors.

7. Color the box for the fifth floor green.

8. Color the box for the twelfth floor blue.

9. Write the number in the top floor.

10. Chloe got on the elevator at the first floor. The elevator went up 7 floors. On what floor was Chloe then?

    _____ floor

11. Brad got on the elevator at the twelfth floor. The elevator went down 6 floors. On what floor was Brad then?

    _____ floor

 12. Write and solve your own elevator problem.

    _____

    _____

    _____

1st floor

Challenge

13. Diego got in the elevator on the first floor. The elevator went up 8 floors. Then it went down 3 floors and then up 5 floors. On what floor was Diego then?

    _____ floor

128 one hundred twenty-eight   CXXVIII   64 + 64

Chapter 6 Lesson 8

Problem Solving Strategy
Look for a Pattern

NCTM Standards 1, 2, 3, 4, 6, 7, 8, 9, 10

Understand
Plan
Solve
Check

1. A rocket travels 10 miles in 2 minutes. The rocket lifts off at 7:00. How far has the rocket traveled by 7:20?   _____ miles

   Explain how you found your answer. _____

2. The first four soccer shirts are numbered 1, 3, 5, and 7. If this pattern continues, what would be the number on the eighth shirt?   _____

   Explain. _____

3. A snail travels 1 foot every 5 minutes. He starts crawling at 9:30. What time will it be when the snail travels 5 feet?   ____ : ____

   Explain. _____

NOTE: Your child is exploring different ways to solve problems. Sometimes using the strategy, *look for a pattern*, is an efficient way to solve a problem.

130 – 1   **CXXIX**   one hundred twenty-nine   129

Problem Solving Test Prep

1. Sean finished baseball practice at 6:00. The practice was one and a half hours long. What time did practice start?

   Ⓐ 4:00

   Ⓑ 4:30

   Ⓒ 5:30

   Ⓓ 7:30

2. Faith made 18 muffins. Ella made 3 less muffins than Faith. How many muffins did they make altogether?

   Ⓐ 15 muffins

   Ⓑ 21 muffins

   Ⓒ 30 muffins

   Ⓓ 33 muffins

 Show What You Know

3. Each tricycle has 3 wheels. How many wheels are there on 6 tricycles?

tricycles	1					
wheels	3					

   _____ wheels

   Explain how you found the answer.

   _____

   _____

   _____

   _____

4. There are 4 children in line. Sam is between Clare and Liddy. Ben is last. Liddy is next to Ben. Who is first in line?

   Explain your answer.

   _____

   _____

   _____

   _____

130 one hundred thirty

Chapter 7

Name _____

Doubling, Halving, and Fractions
Sharing with a Group

You need
- graham crackers
- paper towel

How can you share your snack with others?

STEP 1 Observing

Look at your crackers. Are there enough for you and your partner to share? How do you know?

_____

_____

_____

STEP 2 Sharing

Now work with another partner team. How can you share the new group's crackers equally?

_____

_____

_____

STEP 3 Getting One More

Add one more cracker. Now can you share equally? What can you do?

_____

_____

_____

Investigation

School-Home Connection

Dear Family,

Today we started Chapter 7 of *Think Math!* In this chapter, I will find double and half of a number. I will also identify and write fractions of an object and a set of objects. There are NOTES on the Lesson Activity Book pages to explain what I am learning every day.

Here are some activities for us to do together at home. These activities will help me understand doubles, halves, and fractions.

Love,

_____

Family Fun

Share the Granola Bar!

Work with your child to play this game. Your child will play a similar game in Lesson 7.

- Prepare a gameboard like this. You also need two different-color pencils and a number cube.

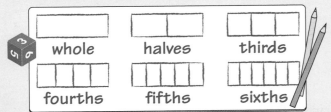

- Take turns with your child. For a turn, toss the number cube. This is the number of people who share the granola bar. Find the bar divided into that many equal pieces, write the fraction in one piece, and color that piece. This shows that you ate it. For example, if you roll a 6, write $\frac{1}{6}$ in one piece of the bar for sixths, and then color it.

- If there are no uncolored pieces left for a fraction, you lose a turn.

- The winner is the last to eat a piece of granola!

Double Your Money

Work with your child to practice doubling a money amount.

- Write a money amount less than 50¢ on a slip of paper. Together, show it with coins.

- Ask your child to double that amount of money using any method. Your child can figure out the amount on a scrap of paper and then show it with coins or figure it out just using the coins.

- As a challenge, you might ask your child to find half of the original amount.

Chapter 7 Lesson 1

Exploring One Half

NCTM Standards 1, 6, 8, 9, 10

Circle the pictures that show one half.

1.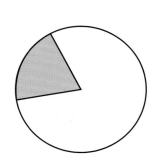

2.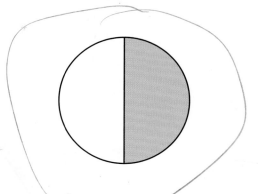

3.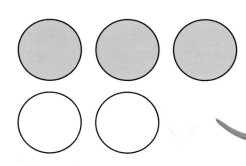

4.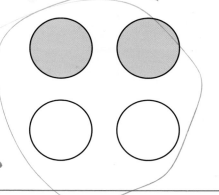

5.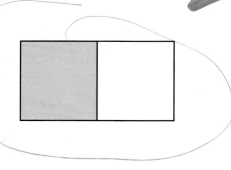

6.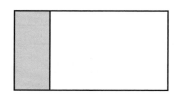

7.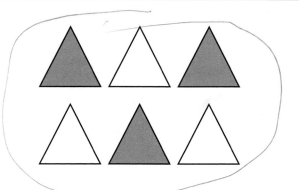

8.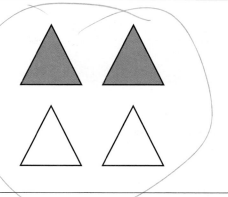

NOTE: Your child is learning to find one half of an object or set of objects.

27   CXXXV   one hundred thirty-five   135

Color $\frac{1}{2}$ of each picture.

9.

10.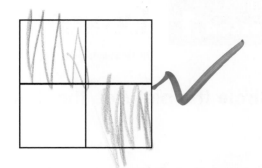

11.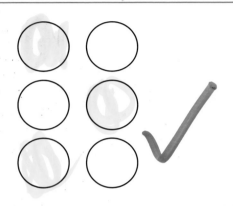

12.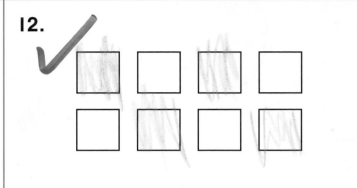

13. Color $\frac{1}{2}$ of the set of circles.
Tell how you know it is $\frac{1}{2}$.

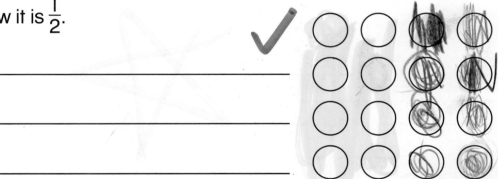

_____

_____

_____

Challenge

14. Dwayne plays video games for half an hour. How many minutes is that? Explain.

_____ minutes

_____

_____

_____

136 one hundred thirty-six CXXXVI △ 68 + 68

Chapter 7 Lesson 2: Finding Half: Even or Odd

NCTM Standards 1, 2, 6, 7, 8, 9, 10

How much is half?

1.
Half of 26 is 13.

2.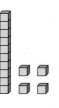
Half of 14 is ___.

3.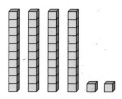
Half of 42 is ___.

4.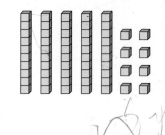
Half of 58 is ___.

	Whole	Half
5.	6	3
6.	8	4
7.	10	0
8.	18	31
9.	24	
10.	36	
11.	70	

NOTE: Your child is learning to find half of even and odd numbers. Together, find half of 20.

What number will solve each riddle?

12. I am an even number. I am half of 44. What number am I?

    22

13. I am an even number. Half of me is 14. What number am I?

    ____

14. I am an even number. Half of me is 36. What number am I?

    ____

15. I am an odd number. I am half of 50. What number am I?

    ____

16. I am an odd number. Half of me is $11\frac{1}{2}$. What number am I?

    ____

17. I am an odd number. Half of me is $9\frac{1}{2}$. What number am I?

    ____

18. Make up your own riddle. Ask a classmate to solve it.

    _____

    _____

    _____

Challenge

Write the same number in frames that are the same.

19. ◯ + ◯ = 86

20. ☐ + ☐ = 56

21. ⬡ + ⬡ = 9

22. △ + △ = 25

138 one hundred thirty-eight   CXXXVIII   138   69 + 69

Chapter 7
Lesson 3
Doubling Numbers

NCTM Standards 1, 2, 6, 7, 8, 9, 10

What is the double of each number?
Draw symbols if you want.

1.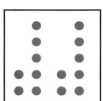

   7 doubled is __14__.

2.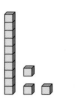

   13 doubled is _____.

3.

   18 doubled is _____.

4.

   20 doubled is _____.

5.

   25 doubled is _____.

6.

   31 doubled is _____.

7.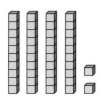

   42 doubled is _____.

8.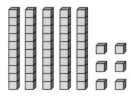

   56 doubled is _____.

NOTE: Your child is learning to double numbers. Together, find the double of 9.

What are the missing numbers? Write each rule.

9.

0	0
1	2
2	4
3	
4	
	10
6	

Rule: _____

10.

5	9
6	11
20	39
3	
10	
11	
	29

Rule: _____

Use doubles to solve.

11. 6 + 6 = _____   6 + 5 = _____   6 + 7 = _____

12. 15 + 15 = _____   15 + 14 = _____   16 + 14 = _____

13. 50 + 50 = _____   49 + 49 = _____   49 + 51 = _____

14. 20 + 20 = _____   19 + 19 = _____   17 + 19 = _____

Problem Solving

15. Kyle has a recipe that makes 12 cups of punch. Kyle doubles the recipe. How many cups will he make?

_____ cups

Kyle doubles the recipe again. How many cups will he have now?

_____ cups

Chapter 7
Lesson 4
Halving and Doubling Time and Numbers
NCTM Standards 1, 2, 6, 7, 8, 9, 10

What is missing?

1.

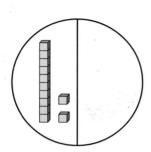

12 is half of __24__.

2.

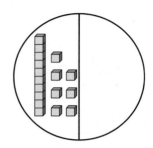

17 is half of _____.

3.

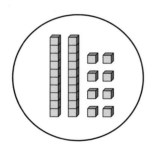

_____ is half of 28.

4.

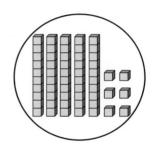

_____ is half of 56.

	Whole	Half
5.		15
6.	46	
7.		36
8.	58	

NOTE: Your child is learning to find doubles and halves.

100 + 40 + 1   **CXLI**   one hundred forty-one   141

How long will a round trip take?

	One Way	Round Trip
9.	9 minutes	18 minutes
10.	15 minutes	_____ minutes
11.	24 minutes	_____ minutes
12.	40 minutes	_____ minutes
13.	55 minutes	_____ minutes

A round trip is double a one-way trip.

 14. How did you find the answer for Problem 13?
Use words, numbers, or pictures to explain.

Challenge

15. What time is the movie half over?
What time will the movie end?

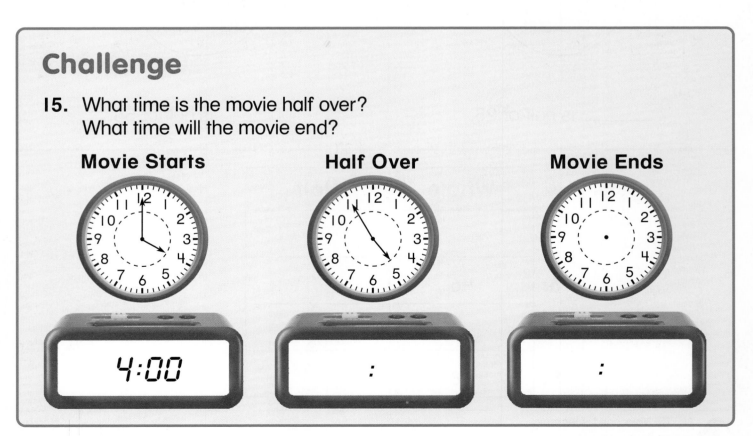

Movie Starts — 4:00
Half Over
Movie Ends

142 one hundred forty-two   CXLII   71 + 71

**Chapter 7
Lesson 5**

Doubling Length

NCTM Standards 1, 2, 3, 4, 6, 8, 9, 10

What is the distance around each figure?

1.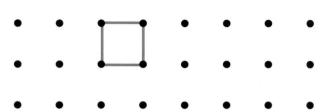

   __4__ spaces

2.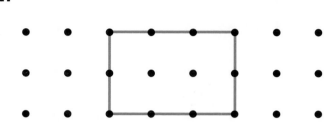

   _____ spaces

3.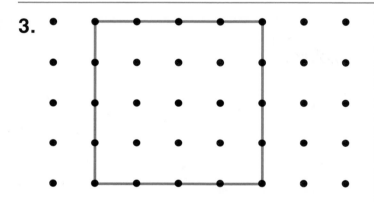

   _____ spaces

4.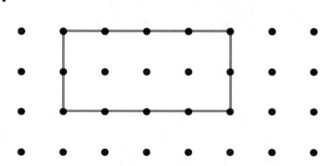

   _____ spaces

5. Draw a figure with 4 sides. What is the distance around the figure?

   _____ spaces

NOTE: Your child is learning to double the length of lines.

11 × 13   **CXLIII**   one hundred forty-three   **143**

Draw a new line. Make it twice as long as the blue line.
How long is each line?

6.

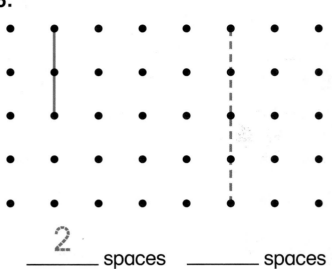

__2__ spaces _____ spaces

7.

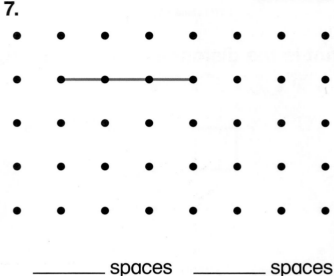

_____ spaces _____ spaces

Draw a new figure. Double the sides of the blue figure.

8.

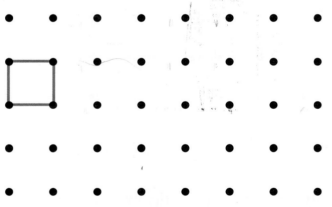

9.

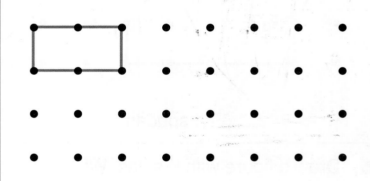

Challenge

10. The distance around a square is 16 spaces. Draw the square. How long is each side?

_____ spaces

144 one hundred forty-four   CXLIV   12 dozen = 1 gross

Chapter 7 Lesson 6

Thirds and Fourths

NCTM Standards 1, 2, 6, 8, 9, 10

Name _____ Date/Time _____

1. Color $\frac{1}{2}$ of the figure.

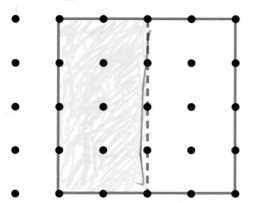

Use the dots to help you divide the whole into equal parts.

Color $\frac{1}{3}$.

2.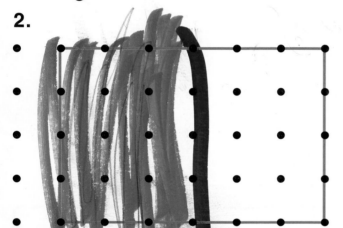

3.

Color $\frac{1}{4}$.

4.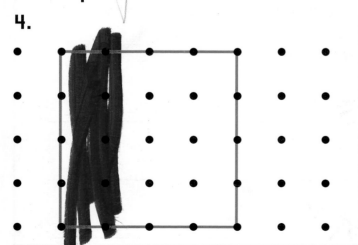

5.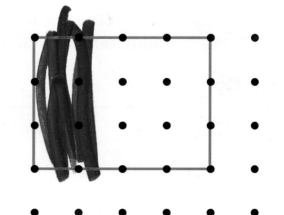

NOTE: Your child is learning to identify and write fractions. Together, fold a napkin to show one third or one fourth.

CXLV   one hundred forty-five   145

What part is colored? Write the fraction.

6. 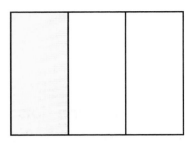 $\frac{1}{3}$

7. 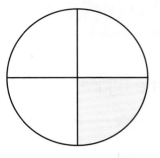 $\frac{1}{4}$

8. 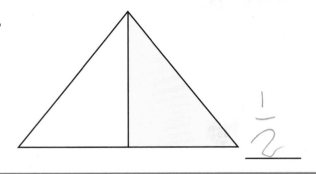 $\frac{1}{2}$

9. 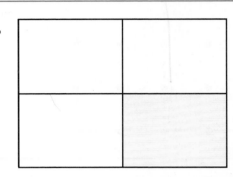 $\frac{4}{\pi}$

Color $\frac{1}{4}$ in three different ways.

10.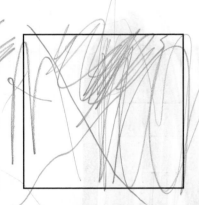

Problem Solving

11. Four friends want to share a whole apple pie. How many pieces do they need to cut the pie into? What fraction of the pie would each person get? Use words, numbers, or pictures to explain.

_____

_____

_____

146 one hundred forty-six CXLVI 73 + 73

Chapter 7
Lesson 7

Fair Shares
NCTM Standards 1, 2, 6, 8, 9, 10

Name _____   Date/Time _____

Write each fraction.

1.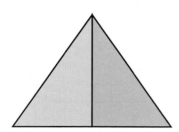

   What part is green? $\frac{1}{2}$

   What part is blue? $\frac{1}{2}$

2.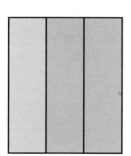

   What part is green? $\frac{1}{3}$

   What part is blue? $\frac{2}{3}$

3.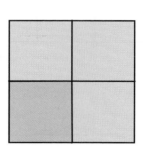

   What part is green? $\frac{1}{4}$

   What part is blue? $\frac{3}{4}$

4.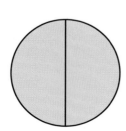

   What part is green? $\frac{2}{2}$

NOTE: Your child is learning to identify and write fractions for parts of a whole.

150 − 3   **CXLVII**   one hundred forty-seven   147

Color to show each fraction.

5. 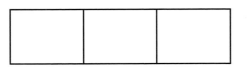 $\frac{2}{3}$

6. 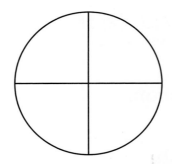 $\frac{3}{4}$

7. 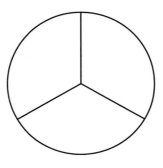 $\frac{3}{3}$

8. 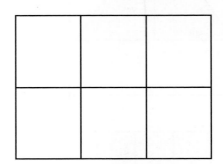 $\frac{4}{6}$

9. Color $\frac{2}{5}$ green. Color $\frac{1}{5}$ blue.

   Color the rest red.

   What fraction of the figure is red? _____

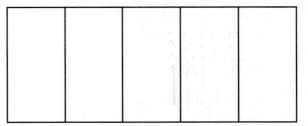

Problem Solving

10. Erin has 4 stuffed elephants. Three of the elephants are pink. What fraction of the elephants is NOT pink? _____

    Draw a picture to show how you found the answer.

Chapter 7 Lesson 8: Exploring Fractions with Cuisenaire® Rods

NCTM Standards 1, 2, 6, 7, 8, 9, 10

The striped rod is one whole. How much is one of the other rods? Write the fraction.

1. One dark green rod is $\frac{1}{2}$.

2. One red rod is _____.

3. One purple rod is _____.

4. One green rod is _____.

5. One white rod is _____.

NOTE: Your child is learning to represent fractions with Cuisenaire® Rods. Ask your child how many green rods make up $\frac{1}{2}$ of a striped rod.

100 + 40 + 9   CXLIX   one hundred forty-nine   149

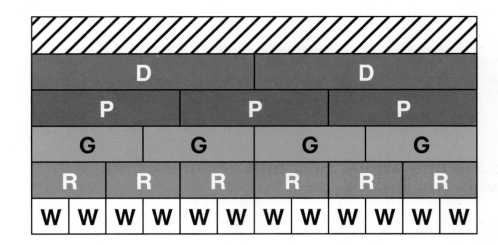

Write <, >, or =. Use the picture to help you.

6. $\frac{1}{3}$ ◯ $\frac{1}{4}$

   P   G

7. $\frac{1}{6}$ ◯ $\frac{1}{4}$

   R   G

8. $\frac{1}{6}$ ◯ $\frac{1}{12}$

9. $\frac{1}{6}$ ◯ $\frac{2}{12}$

10. Make your own. Choose a rod in two different colors. Complete the sentence.

Challenge

11. $\frac{1}{4} + \frac{1}{4} + \frac{1}{4}$ ◯ $\frac{1}{3} + \frac{1}{3}$

12. $\frac{1}{3}$ ◯ $\frac{1}{6} + \frac{1}{6}$

Chapter 7
Lesson 9: More Fractions

NCTM Standards 1, 2, 6, 7, 8, 9, 10

What part is colored? Write the fraction.

1. 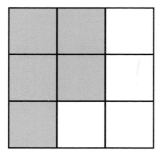 $\frac{5}{9}$

2. 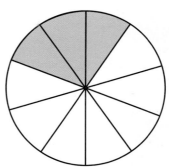 _____

3. 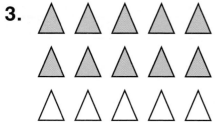 _____

4. 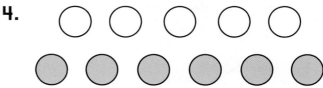 _____

5. 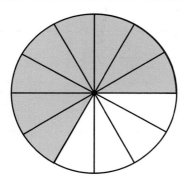 _____

6. 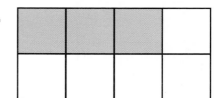 _____

**Make your own. Color part of the picture.
Write the fraction you made.**

7. 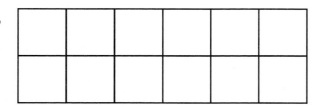 _____

8.  _____

 NOTE: Your child is learning to write fractions and estimate the relative size of fractions.

Write the fraction for the colored part.
Circle if it is closer to 0, $\frac{1}{2}$, or 1.

9.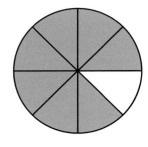

$\frac{7}{8}$

| 0 | $\frac{1}{2}$ | (1) |

10.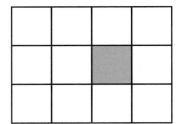

_____

| 0 | $\frac{1}{2}$ | 1 |

11.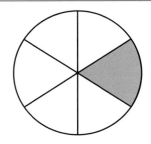

_____

| 0 | $\frac{1}{2}$ | 1 |

12.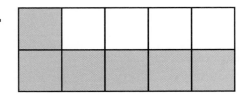

_____

| 0 | $\frac{1}{2}$ | 1 |

Problem Solving

13. Color the picture to show a fraction close to $\frac{1}{2}$.

   What fraction did you show? _____

   Explain how you know it is close to $\frac{1}{2}$.

   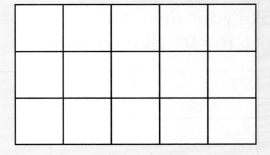

   _____

   _____

152 one hundred fifty-two   CLII   76 + 76

Chapter 7
Lesson 10

Problem Solving Strategy
Guess and Check

NCTM Standards 1, 2, 4, 5, 6, 7, 8, 9, 10

Understand
Plan
Solve
Check

1. Dan has a cup of marbles. He takes $\frac{1}{2}$ and shares the rest fairly between 2 friends. Dan gets 2 more marbles than each friend. How many marbles were in the cup?

   _____ marbles

   How did you find the answer? _____

   _____

   _____

   _____

2. Double a number and it is half of 48.
   What is the number? Explain. _____

   _____

   _____

   _____

3. There are fewer than 10 pretzels in the bag. If two children share, 1 pretzel is left over. If three children share, 2 pretzels are left over. How many pretzels are in the bag?

   _____ pretzels

   How did you find the answer? _____

   _____

   _____

 NOTE: Your child is exploring different ways to solve problems. Sometimes using the strategy, *guess and check*, is an efficient way to solve a problem.

100 + 50 + 3      CLIII   one hundred fifty-three   153

Problem Solving Test Prep

1. Maria has pennies and dimes in a bag. She picks 3 coins. Which is NOT an amount of money she could have?

   Ⓐ 3¢

   Ⓑ 12¢

   Ⓒ 15¢

   Ⓓ 21¢

2. Sal's team scored 21 points in the second half. They had 50 points at the end of the game. How many points did the team get in the first half?

   Ⓐ 21 points

   Ⓑ 29 points

   Ⓒ 39 points

   Ⓓ 71 points

 Show What You Know

3. Nicole rides her bike 1 mile every 10 minutes. She starts riding at 9:00. What time will it be after she rides 4 miles?

   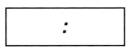

Miles					
Time	:	:	:	:	:

   Explain your answer.

   _____

   _____

   _____

   _____

4. A pizza is cut in 8 pieces. The children eat half a pizza. How many pieces are left?

   _____ pieces

   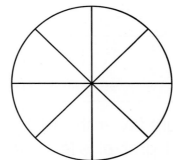

   Explain your answer.

   _____

   _____

   _____

Chapter 7 Review/Assessment

NCTM Standards 1, 2, 4, 6, 7, 8, 9, 10

Color $\frac{1}{2}$ of each picture. Lesson 1

1.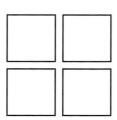

2.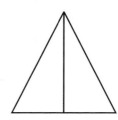

How much is half of each amount? Lesson 2

3.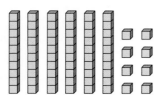

Half of 68 is _____.

4.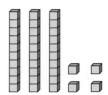

Half of 34 is _____.

How much is double of each amount? Lesson 3

5.

25 doubled is _____.

6.

40 doubled is _____.

7. How long will a round trip take? Lesson 4

One Way	12 minutes	46 minutes	34 minutes
Round Trip	_____ minutes	_____ minutes	_____ minutes

CLV one hundred fifty-five

Draw a new figure. Double the sides of the blue figure. Lesson 5

8.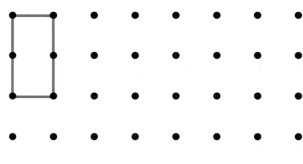

9.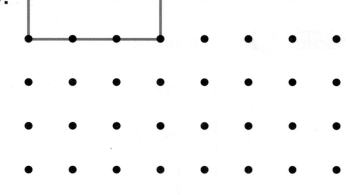

What part is green? Write each fraction. Lessons 6, 7

10.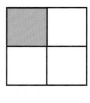

    _____

11.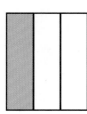

    _____

12.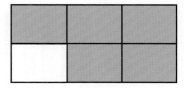

    _____

13. Write the fraction for the colored part. Circle if it is closer to 0, $\frac{1}{2}$, or 1. Lesson 9

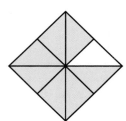

_____

| 0 | $\frac{1}{2}$ | 1 |

Problem Solving Lesson 10

14. Jess has a bag of cherries. She eats half. Then she gives her brother half of what is left. She is left with 5 cherries. How many cherries did she have to start?

_____ cherries

156 one hundred fifty-six　CLVI　13 dozen

Chapter 8

Name _____

Building Addition and Subtraction Fluency
Exploring Related Number Sentences

You need
- 8 index cards

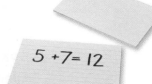

How can two number sentences be related?

STEP 1 Observing

Turn over two cards from the pile. Are the number sentences related? Explain.

_____
_____
_____

STEP 2 Observing Some More

Pick another card, and put it with the other two. Do you see any related sentences now? Explain.

_____
_____
_____

STEP 3 Writing Related Sentences

Pick one of your cards. Write a related number sentence. How are they related?

_____
_____
_____

School-Home Connection

Dear Family,

Today we started Chapter 8 of *Think Math!* In this chapter, I will develop my ability to add and subtract bigger numbers. There are NOTES on the Lesson Activity Book pages to explain what I am learning every day.

Here are some activities for us to do together at home. These activities will help me understand addition and subtraction.

Love,

_____

Family Fun

Race to 100!

Play this game with your child. Your child will play this game in Lesson 6.

- Prepare two gameboards like this. You also need two number cubes.

I already have...	The sum of the numbers tossed is...	My new total is...
0		

- Players alternate turns. For each turn, toss the number cubes and add the two numbers. Record the sum in the middle column. Add the first two columns together and record that sum in the third column. Copy the "new total" into the first column of the row for the next turn.
- The winner is the first player to reach 100 in the "new total" column.

What Makes a Number

Work with your child to find addition problems for a given number.

- On index cards or slips of paper, write a variety of two-digit numbers, one per card.
- Place the cards face down in a pile, and have your child choose one card. Together, name two numbers that when added together make the number on the card. Write an addition sentence for your numbers.

51   25 + 26 = 51

- Repeat, reusing cards if you want. If cards are reused, challenge your child to come up with a different addition sentence for the number.

Chapter 8 Lesson 1: Adding with Cuisenaire® Rods

NCTM Standards 1, 2, 4, 6, 7, 8, 9, 10

What are all the possible 2-car trains?

1. [R]
   W ___

2. [G]
   R ___
   ___ ___

3. [P]
   R ___
   ___ ___
   ___ ___

What are all the possible 3-car trains?

4. [G]
   W ___ ___

5. [P]
   R ___ ___
   ___ ___ ___
   ___ ___ ___

6. [Y]
   ___ ___ ___
   ___ ___ ___
   ___ ___ ___
   ___ ___ ___
   ___ ___ ___
   ___ ___ ___

7. How many different rod trains can you make?

Rod Color	2-Car Trains
red	1
green	
purple	

Rod Color	3-Car Trains
green	
purple	
yellow	

NOTE: Your child is learning to list combinations and write addition sentences using Cuisenaire® Rods.

Write an addition sentence for each picture.

8.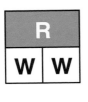

W + W = R

9.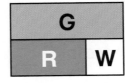

_____

10.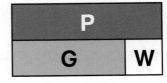

_____

11.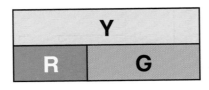

_____

12.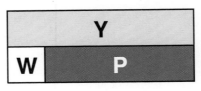

_____

13. What other rod train is as long as a yellow rod? Write an addition sentence and draw a picture.

_____

Challenge

14. Make a 2-car train that is as long as a yellow rod. Then make a new train using two of each car. What color rod matches this longer train? _____

What addition sentence shows this? _____

160 one hundred sixty   CLX   32

Chapter 8 Lesson 2: Exploring Fact Families

NCTM Standards 1, 2, 6, 7, 8, 9, 10

What is the fact family for each picture?

1.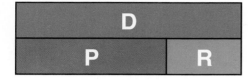

   R + P = ___
   ___ + ___ = ___
   ___ − ___ = ___
   ___ − ___ = ___

2.

   ___ + ___ = ___
   ___ + ___ = ___
   ___ − ___ = ___
   ___ − ___ = ___

3.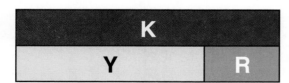

   ___ + ___ = ___
   ___ + ___ = ___
   ___ − ___ = ___
   ___ − ___ = ___

NOTE: Your child is learning to write a fact family of addition and subtraction sentences for Cuisenaire® Rod trains.

100 + 50 + 11   **CLXI**   one hundred sixty-one

4. What is the fact family for these rod trains?

_____ + _____ = _____

_____ + _____ = _____

_____ − _____ = _____

_____ − _____ = _____

---

5. Write the other sentences for this fact family.
Tell how you completed the sentences.

R + G = Y   _____ − _____ = _____

_____ + _____ = _____   _____ − _____ = _____

_____

_____

_____

Challenge

6. Use ☾ + ☺ = △ to complete each sentence.

☺ + ☾ = _____   △ − ☺ = _____

△ − ☾ = _____

162   one hundred sixty-two   CLXII   81 + 81

Chapter 8 Lesson 3: Connecting Addition and Subtraction

NCTM Standards 1, 2, 6, 7, 8, 9, 10

**What is missing in each fact family?
Draw the symbols. Write the numbers.**

1.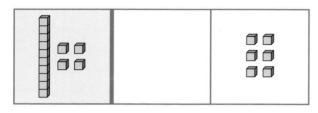

   ___ 14 ___ − ___ = ___ 6 ___

   ___ − ___ = ___

   ___ + ___ = ___

   ___ + ___ = ___

2.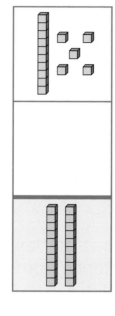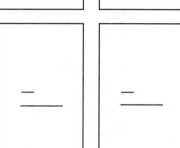

3.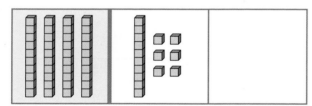

   ___ = ___ + ___

   ___ = ___ + ___

   ___ = ___ − ___

   ___ = ___ − ___

4.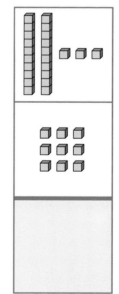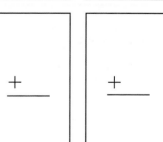

NOTE: Your child is learning to write fact families that relate addition and subtraction problems.

90 + 70 + 3   CLXIII   one hundred sixty-three

What is missing in each fact family?
Draw the symbols. Write the numbers.

5.

6.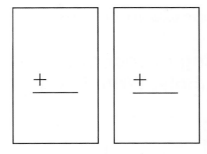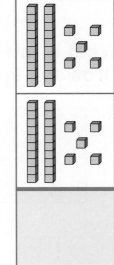

_____ − _____ = _____

_____ − _____ = _____

_____ + _____ = _____

_____ + _____ = _____

7. Pick a number sentence from Problem 6.
   Write a story to match it.

_____

_____

_____

_____

Challenge

8. Write the fact family.

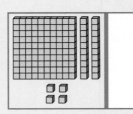

_____ = _____ + _____   _____ = _____ − _____

_____ = _____ + _____   _____ = _____ − _____

Chapter 8 Lesson 4

Adding and Subtracting Using 5 and 10

NCTM Standards 1, 6, 7, 8, 9, 10

How much is each collection worth?

1.

   _____ ¢

2.

   _____ ¢

What is missing? Complete the expressions for each number.

3. 8

   5 + __3__

   4 + __4__

   10 − __2__

4. 14

   5 + _____

   7 + _____

   10 + _____

5. 15

   5 + _____

   5 + 5 + _____

   20 − _____

6. 21

   20 + _____

   1 + 10 + _____

   10 + _____

7. 25

   5 + _____

   10 + 10 + _____

   15 + _____

8. 36

   26 + _____

   30 + _____

   16 + _____

NOTE: Your child is learning a strategy to help add and subtract by breaking numbers apart into fives and tens.

Add or subtract.

9. $8 + 21 =$ _____

10. $14 + 9 =$ _____

11. $15 + 25 =$ _____

12. $14 - 6 =$ _____

13. $21 - 8 =$ _____

14. $36 - 25 =$ _____

15.
```
   27¢
 +  5¢
 ─────
    ¢
```

16.
```
   40¢
 - 15¢
 ─────
    ¢
```

17.
```
   25¢
 -  9¢
 ─────
    ¢
```

 18. Use words, numbers, or pictures to explain how you solved Problem 17.

Problem Solving

19. Barbara has 35¢. She buys a whistle for 10¢ and a sticker for 8¢. How much money does she have left? Explain. _____¢

Chapter 8 Lesson 5: Adding and Subtracting Numbers Near 10

NCTM Standards 1, 2, 6, 7, 8, 9, 10

What is missing in each table? Use the rule.

1.

a	11	12	21	22	30	**31**	32	43	**44**
$a + 8$	19	20	29	**30**	**38**	39	**40**	**51**	52

2.

n	29	28	50	51	**52**	61	**71**	91	**92**
$n - 9$	20	19	**41**	**42**	43	**52**	62	**82**	83

3.

10	20	**30**	40	90	**91**	92	73	**74**	c
2	**12**	22	**32**	**82**	83	**84**	**65**	66	$c - 8$

4.

28	27	78	77	58	57	37	**17**	87	m
37	**36**	87	**86**	**67**	**66**	**46**	26	**96**	$m + 9$

NOTE: Your child is learning to solve addition and subtraction problems by using 10 and then adjusting the results.

What is missing? Complete each sentence.

5. 37 + 8

    37 + 10 = 47

    47 − 2 = 45

6. 28 + 9 = _____

7. 12 + 49 = _____

8. 64 + 11 = _____

Find convenient numbers to help you add and subtract.

9. 42 − 8

    42 − 10 = ☐

    ☐ + 2 = _____

10. 33 − 9 = _____

11. 51 − 12 = _____

12. 25 − 11 = _____

  13. How would you solve 54 − 9? Use words, numbers, or pictures to explain.

Challenge

14. Complete the table in two different ways. Write the rules.

a	6	15	23	47	58
	12				

a	6	15	23	47	58
	12				

Place Value and Cross Number Puzzles

Chapter 8, Lesson 6

NCTM Standards 1, 2, 6, 8, 9, 10

What is missing? Write the numbers or the symbols.

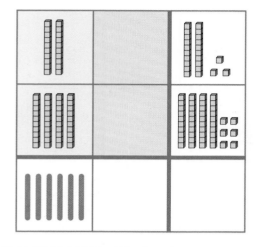

1.

2.

10		17
20	5	25

3.

4.

	4	
10		16
20		

5.

20		21
		45
	3	13
70	9	

NOTE: Your child is learning to add and subtract numbers using Cross Number Puzzles. Ask your child to explain how to solve Problem 5.

What is missing? Write the numbers.

6.

	0	30
0		9
	8	18
40		

7.

	7	
0	6	
40		42
80	15	

8.

10	5	
	4	
20		28
80		

9.

31		1
	10	8
29		
		18

Challenge

10. Fill in the Cross Number Puzzle to help solve this problem.

```
   86
 + 47
```

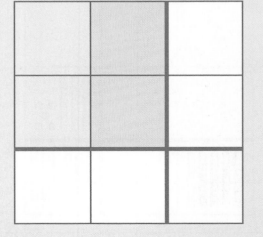

170 one hundred seventy   CLXX   34

Chapter 8 Lesson 7: Breaking Numbers Apart

NCTM Standards 1, 2, 6, 7, 8, 9, 10

5 + 3, 1 + 7, and 10 − 2 are all different ways to show 8.

Break the numbers in different ways so they are easier to add. Write the missing numbers.

1.
| 8
 + 7 | ⇨ | 5 + 3
 + 5 + 2
 ――――
 + 5 | ⇨ | 1 + 7
 + 4 + 3
 ――――
 + | ⇨ | 10 − 2
 + 10 − 3
 ――――
 − |

2.
| 15
 + 18 | ⇨ | 10 + 5
 + 10 + 8
 ――――
 + | ⇨ | + 10
 + 5 +
 ――――
 + | ⇨ | 20 −
 +   − 2
 ――――
 − 7 |

3.
| 9
 + 34 | ⇨ | 4 +
 + 9 +
 ――――
 13 + | ⇨ | 9 +
 + 31 +
 ――――
 + | ⇨ | 10 −
 +   − 6
 ――――
 − |

4.
| 26
 + 26 | ⇨ | + 1
 +   + 1
 ――――
 + 2 | ⇨ | 20 +
 + 20 +
 ――――
 + | ⇨ | 30 −
 +   − 4
 ――――
 − |

NOTE: Your child is learning strategies that make it easier to solve addition and subtraction problems.

100 + 60 + 11   △   **CLXXI**   one hundred seventy-one   **171**

Break the numbers apart in different ways. Then complete the puzzle. Circle the way that matches the puzzle.

5.
14		+ 4		7 +		20 −
+ 28	⇒	+ 20 + ___	⇒	+ 14 + ___	⇒	+ ___ − 2
		+		+		− 8

Puzzle:
10		14
	8	28
	12	

6.
76		75 +		+ 6		80 −
+ 28	⇒	+ 25 + ___	⇒	+ 20 + ___	⇒	+ ___ − 2
		+		+		−

Puzzle:
70		
	8	
90		

7.
37		20 +		40 −		+ 7
+ 19	⇒	+ 16 + ___	⇒	+ ___ − 1	⇒	+ 10 + ___
		+		−		+

Puzzle:
30	7	
10		19

Challenge

8. Break the numbers apart in different ways.

149		+ ___ + 100		150 −
+ 46	⇒	+ 6 + ___ + 0	⇒	+ ___ − 4
		+ ___ + ___		−

172 one hundred seventy-two   CLXXII   86 + 86

Chapter 8
Lesson 8

Using Cross Number Puzzles to Subtract

NCTM Standards 1, 2, 6, 7, 8, 9, 10

What is missing?

1.

40		52
20	4	24

2.

71	60	11
46		6

3.

76		
	80	13

4.

	9	59
70		83

Show how to solve each problem.

5. 82
    -69

6. 38
    $+47$

NOTE: Your child is learning to solve addition and subtraction problems using different strategies.

What is missing?

7. 26
    + 46
    ———

8. 62
    − 30
    ———

Think about the numbers in a problem. What method would work the best?

9. 34 − 19 = _____

10. 51 + 25 = _____

11. 98 + 35 = _____

12. 196 − 97 = _____

 13. How would you solve 53 − 29? Use words, numbers, or pictures to explain.

Problem Solving

14. Lenny read 49 pages of his book on Monday. He read 23 pages on Tuesday. How many pages did he read on both days?

    _____ pages

15. Pia's book is 302 pages long. She has read 150 pages. How many more pages does Pia have left to read?

    _____ pages

Chapter 8 Lesson 9

Comparing Mathematical Expressions

NCTM Standards 1, 2, 6, 7, 8, 9, 10

Make each sentence true. Write >, <, or =.

1. 64 + 36 ◯< 64 + 37
2. 54 − 5 ◯ 54 − 10
3. 84 + 37 ◯ 37 + 84
4. 87 − 48 ◯ 87 − 50
5. 72 + 43 ◯ 42 + 72
6. 39 + 12 ◯ 13 + 39

7. 36 − 26 ◯ 86 − 76
8. 57 − 37 ◯ 46 − 36
9. 24 + 36 ◯ 25 + 35
10. 64 − 39 ◯ 65 − 39
11. 64 − 39 ◯ 65 − 40
12. 39 + 13 ◯ 12 + 40

NOTE: Your child is learning to compare addition and subtraction expressions using <, >, and =.

CLXXV   one hundred seventy-five   175

What is missing?

13. 28 + 32 = __32__ + 28

14. 43 − _____ = 43 − 19

15. 15 + 17 = 16 + _____

16. _____ + 63 = 12 + 64

17. 30 + 28 = 20 + _____

18. 47 − 18 = 57 − _____

19. 24 − 12 = _____ − 13

20. 62 + 36 = _____ + 60

 21. How did you solve Problem 20? Use words, numbers, or pictures to explain.

Challenge
Write >, <, or =.

22. n + 9 ◯ n + 10

23. n − 36 ◯ n − 37

176 one hundred seventy-six   CLXXVI   88 + 88

Chapter 8 Lesson 10
Creating and Solving Story Problems

NCTM Standards 1, 2, 6, 7, 8, 9, 10

> Jamie sold 63 red balloons, 48 blue balloons, and 24 green balloons.

1. Circle a question that can be answered from the story.

   • How many balloons did Jamie sell?

   • How much money did Jamie make?

   • How many more red balloons than blue balloons did Jamie sell?

   • How much string did Jamie use?

2. Write a number sentence to match the question you circled.

 3. Show how you would solve the problem.

NOTE: Your child is learning to work with addition and subtraction story problems. Ask him or her to make up a story problem for you to solve.

59 + 59 + 59   CLXXVII   one hundred seventy-seven   177

Which number sentence matches each story?

4. Ms. Lee buys 73 notebooks. Mr. Hall buys the same number of notebooks.

$73 + 27 = \underline{}$

5. Andrew collected 73 shells. He used 27 of the shells to make a frame. How many shells are not in the frame?

$73 = \underline{}$

6. Tamika had 73 crayons in her box. She found 27 more. Now how many crayons does she have?

$73 - 27 = \underline{}$

Challenge

7. Solve Problem 6. Show how you solved it.

Chapter 8 Lesson 11: Strategies for Multiple-Choice Questions

NCTM Standards 1, 6, 8, 9, 10

Name _____ Date/Time _____

Fill in the bubble for each correct answer.

It helps to cross out answers that do not make sense for a problem.

Color in the correct answer.

1. 13
   + 17

   Ⓐ 4   ● 30
   Ⓑ 20   Ⓓ 45

2. 3
   + 7

   Ⓐ 4
   Ⓑ 8
   Ⓒ 9
   Ⓓ 10

3. 30 − 10 = _____

   Ⓐ 20   Ⓒ 40
   Ⓑ 31   Ⓓ 300

4. 43
   − 28

   Ⓐ 71
   Ⓑ 61
   Ⓒ 25
   Ⓓ 15

5. 25 + 17 = _____

   Ⓐ 8   Ⓒ 42
   Ⓑ 32   Ⓓ 312

NOTE: Your child is learning to answer multiple-choice questions. Ask him or her how to eliminate answer choices to help solve the problems.

100 + 70 + 9   **CLXXIX**   one hundred seventy-nine   179

Fill in the bubble for each correct answer.

6. 49
    + 97

    Ⓐ 1,316
    Ⓑ 154
    Ⓒ 146
    Ⓓ 50

7. 56
    − 19

    Ⓐ 37
    Ⓑ 40
    Ⓒ 43
    Ⓓ 75

8. 175 + 20 = _____

    Ⓐ 195
    Ⓑ 177
    Ⓒ 165
    Ⓓ 155

9. 315 + 9 = _____

    Ⓐ 325
    Ⓑ 324
    Ⓒ 306
    Ⓓ 225

BIG TOY SALE!
trains 12¢   cars 35¢   boats 25¢

10. Emma bought a toy car. She paid the clerk 50¢. How much change did she get?

    Ⓐ 15¢
    Ⓑ 25¢
    Ⓒ 38¢
    Ⓓ 40¢

11. Dane bought a train. Ciara bought a boat. How much more did Ciara spend than Dane?

    Ⓐ 9¢
    Ⓑ 13¢
    Ⓒ 23¢
    Ⓓ 37¢

Challenge

12. Brad bought 3 toy cars and 1 train. Claudia bought 6 boats. How much more did Claudia spend than Brad?

    Ⓐ 47¢
    Ⓑ 45¢
    Ⓒ 43¢
    Ⓓ 33¢

**Chapter 8
Lesson 12**

Problem Solving Strategy
Solve a Simpler Problem

NCTM Standards 1, 2, 6, 7, 8, 9, 10

**Understand
Plan
Solve
Check**

1. Martha had 57 stamps in her collection. She bought a package of 15 stamps. Then how many stamps did she have in the collection?

   _____ stamps

   How did you find the answer?

   _____

   _____

2. There were 184 magazines in the store. Another 55 magazines were delivered. Then the store sold 50 magazines. How many magazines were in the store then?

   _____ magazines

   How did you find the answer?

   _____

   _____

3. Howie spent 16 minutes on his math homework and 25 minutes on his reading homework. How much time did Howie spend doing his homework?

   _____ minutes

   How did you find the answer?

   _____

   _____

 NOTE: Your child is exploring different ways to solve problems. Sometimes solving a simpler problem is an efficient way to find the answer.

100 + 70 + 11      CLXXXI   one hundred eighty-one

Problem Solving Test Prep

1. Kyle rides his bike 1 mile every 10 minutes. He starts riding at 9:00. How far will he ride by 9:30?

    Ⓐ 1 mile

    Ⓑ 2 miles

    Ⓒ 3 miles

    Ⓓ 4 miles

2. Half of a number is between 20 and 25. What number could it be?

    Ⓐ 11

    Ⓑ 24

    Ⓒ 46

    Ⓓ 55

 Show What You Know

3. Tina makes a bead necklace. She strings on 1 white bead, 2 blue beads, 1 white bead, 2 blue beads. If she continues this pattern, what color will the eighth bead be?

    _____

    Explain how you found the answer.

    _____

    _____

    _____

4. The Eagle basketball team scored 16 points in the second half of the game. They had 38 points at the end of the game. How many points did the team score in the first half?

    _____ points

    Explain how you know.

    _____

    _____

    _____

Chapter 8 Review/Assessment

NCTM Standards 1, 2, 6, 8, 9, 10

What is missing in each fact family? Draw the symbols or write the numbers. Lessons 1–3

1.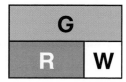

   ___ + ___ = ___

   ___ + ___ = ___

   ___ − ___ = ___

   ___ − ___ = ___

2.

   ___ − ___ = ___

   ___ − ___ = ___

   ___ + ___ = ___

   ___ + ___ = ___

What is missing? Complete the expressions for each number. Lesson 4

3. | 12 |

   10 + ___

   6 + ___

   5 + ___

4. | 34 |

   30 + ___

   24 + ___

   14 + ___

5. **What is missing in the table? Use the rule.** Lesson 5

m	13	14	50		71	72	38		49
$m + 11$	24		61	60				50	

61 + 61 + 61   CLXXXIII   one hundred eighty-three   183

What is missing? Complete each puzzle. Lessons 6–8

6.

	2	
40		45
70		

7.

60	11	71
	3	53

8. Break the numbers apart. Circle the box that matches the puzzle. Lesson 7

$$\begin{array}{r}24\\+19\\\hline\end{array}$$ ⇒ $$\begin{array}{r}+\ 4\\+10+\\\hline\\+\end{array}$$ ⇒ $$\begin{array}{r}+\ 1\\+10+\\\hline\\+\end{array}$$ ⇒ $$\begin{array}{r}30-\\+\quad-1\\\hline\\-\end{array}$$

20		
	9	19
	13	

Make each sentence true. Write >, <, or =. Lesson 9

9. 21 + 38 ◯ 20 + 38

10. 45 − 13 ◯ 45 − 23

Fill in the bubble for each correct answer. Lesson 11

11. $$\begin{array}{r}39\\+\ 25\\\hline\end{array}$$

Ⓐ 514
Ⓑ 64
Ⓒ 54
Ⓓ 14

12. $$\begin{array}{r}87\\-\ 38\\\hline\end{array}$$

Ⓐ 125
Ⓑ 59
Ⓒ 49
Ⓓ 45

Problem Solving Lesson 12

13. Ms. Chen sold 23 apple muffins at the bake sale. She sold 41 corn muffins. How many corn and apple muffins did she sell?

_____ muffins

184 one hundred eighty-four CLXXXIV 92 + 92

Name _____

Chapter 9
Two-Dimensional Figures and Spatial Sense
The Shape of Signs

You need
- art paper
- colored pencils or markers
- straightedge

Create and sort signs.

STEP 1 Making a Sign

Draw a sign on a sheet of paper.

How many sides does your sign have? _____

How many corners does your sign have? _____

STEP 2 Sorting the Signs

Sort the signs in any way. Use words to tell how you sorted.

_____

_____

STEP 3 Sorting Another Way

Tell another way you could sort the signs.

_____

_____

_____

_____

_____

Investigation

School-Home Connection

Dear Family,
Today we started Chapter 9 of *Think Math!* In this chapter, I will explore two-dimensional figures, symmetry, area, and how to record paths. There are NOTES on the Lesson Activity Book pages to explain what I am learning every day.

Here are some activities for us to do together at home. These activities will help me understand two-dimensional figures and symmetry.

Love,

_____

Family Fun

I Spy

Work with your child to play a game about two-dimensional figures, called *I Spy*.

- Review the names of these figures.

square   rectangle   triangle

- To play the game, you secretly choose an object in the room that is (or contains) one of these figures. Then you say, "I spy a square (rectangle or triangle)."

- Your child asks *yes/no* questions about the object until he or she guesses the correct one. Some good questions to ask are: "Is the square in a place that we can easily see?" or "Does the object have more than one square?"

- Take turns selecting an object and asking the questions.

Lines of Symmetry

Work with your child to make a figure that has a line of symmetry.

- You will need paper and scissors. Help your child fold the sheet of paper in half. Make sure the two halves match.

- Help your child draw a design that starts and ends at the fold. Together, guess what the shape will look like when it is cut out and the paper is opened.

- Hold the folded side of the paper and cut out the design. You should not cut along the fold.

- Invite your child to unfold the paper and draw a line down the fold. You made a line of symmetry!

Chapter 9 Lesson 1

Sorting Polygons by Attributes

NCTM Standards 2, 3, 6, 7, 8, 9, 10

Name _____ Date/Time _____

Count the polygons in the picture. Complete each sentence.

1. There are __18__ polygons in the picture.

2. There is _____ polygon with 3 sides in the picture.

3. There are _____ polygons with 4 sides in the picture.

4. There are _____ polygons with more than 4 sides.

 NOTE: Your child is learning about polygons. A polygon is a closed figure with all straight sides.

100 + 80 + 7   **CLXXXVII**   one hundred eighty-seven   187

5. Draw a design using only polygons. Use a straightedge. Start and end each line at a dot.

Problem Solving

6. I have 4 straight sides. I am closed and have no curves. All of my sides are the same length. What figure am I?

_____

Chapter 9 Lesson 2: Congruent and Similar Figures

NCTM Standards 3, 4, 6, 7, 8, 9, 10

1. Draw the same polygon in different positions. Use 4 squares for each polygon. Draw as many congruent polygons as you can.

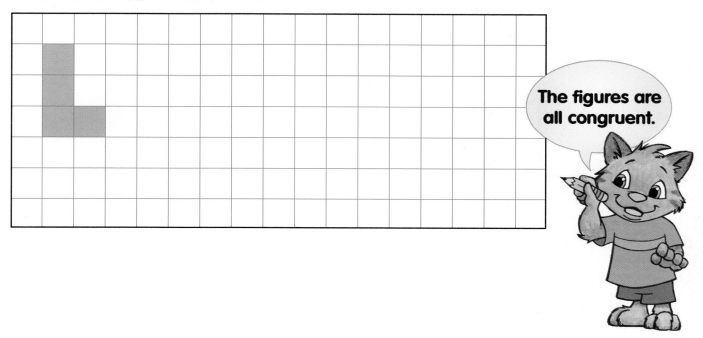

The figures are all congruent.

2. Draw a figure congruent to this one.

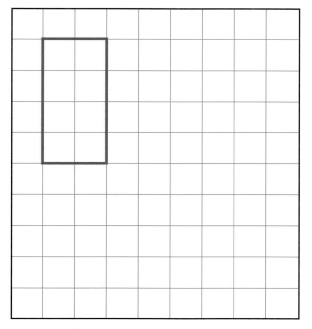

3. Draw 2 congruent figures of your own.

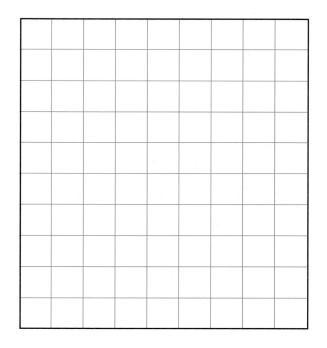

NOTE: Your child is learning about congruent and similar figures. Congruent figures are the same size and shape. Similar figures may be different sizes.

100 + 80 + 9   CLXXXIX   one hundred eighty-nine   189

**Circle the pairs of figures that are similar.
Put an X on the pairs that are *not* similar.**

4.
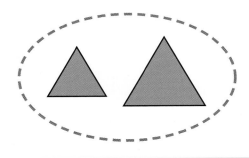

5.
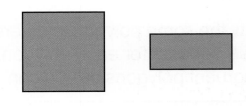

6.
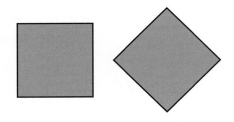

7.
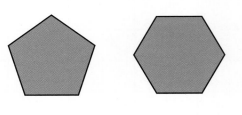

8. Draw 2 figures similar to this one.

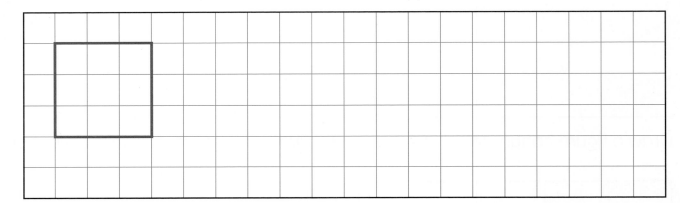

Problem Solving

9. Robin says that she can make a square with 5 congruent squares. Is she correct?
Use words, numbers, or pictures to explain.

Chapter 9
Lesson 3

Building with Triangles

NCTM Standards 2, 3, 4, 6, 7, 8, 9, 10

Draw lines in each figure to make 4 small congruent triangles.

Congruent triangles are the same size and shape.

1.

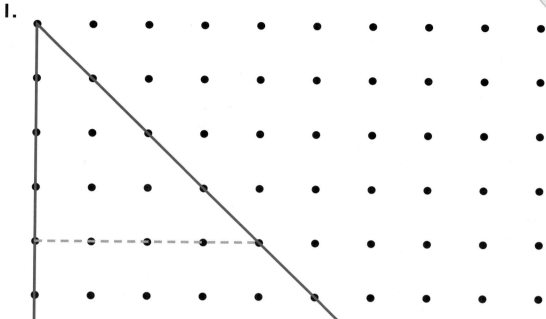

2.

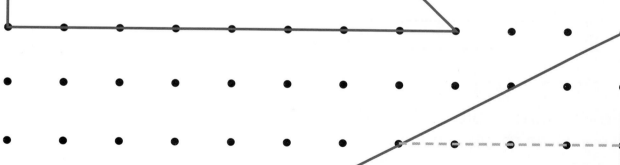

NOTE: Your child is learning to make congruent and similar triangles. Similar triangles have the same shape, but do not have to be the same size.

100 + 80 + 11   **CXCI**   one hundred ninety-one   191

3. Draw a similar triangle on the grid.

 4. What is the same about these 2 triangles? What is different?

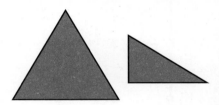

_____

_____

_____

_____

Challenge

5. Draw a similar figure on the grid. Use 4 of the smaller figures.

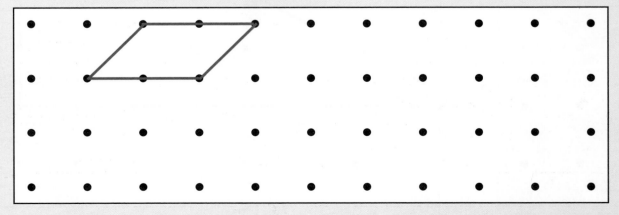

Chapter 9
Lesson 4 — Looking at Reflections

NCTM Standards 3, 6, 7, 8, 9, 10

Draw the reflection of each figure. Use a mirror to help you.

The thick green line shows where to place the mirror.

1.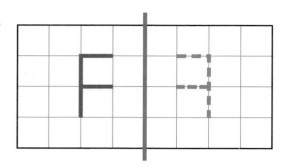

2.

3.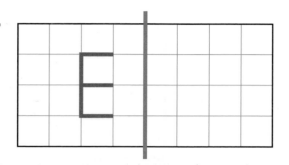

4.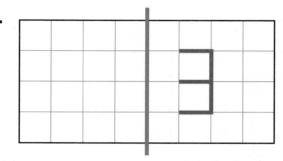

5.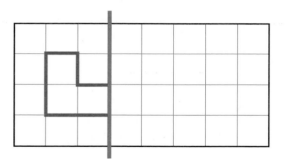

6.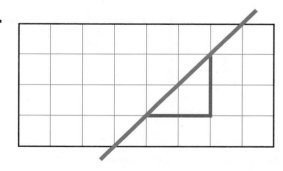

7.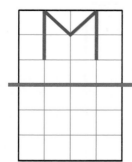

8.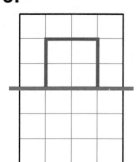

9.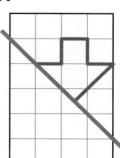

10.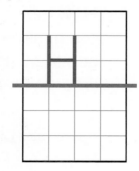

NOTE: Your child is learning to draw reflections of letters and figures. Together, hold a page from a book in front of a mirror and try to read the words.

100 + 80 + 13   **CXCIII**   one hundred ninety-three   193

Draw the reflections. Work across and down within each grid.

11.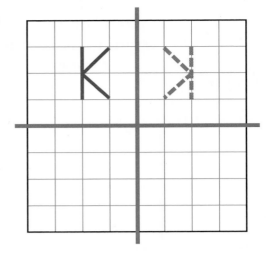

12.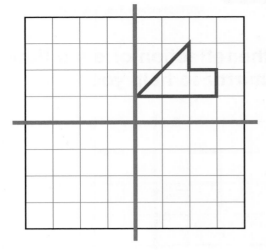

13.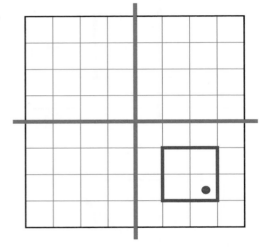

14.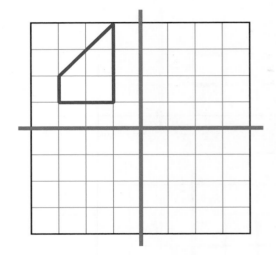

Challenge

15. How many different places can you put a mirror on the square and see the whole square? Draw a line where you put the mirror each time.

    _____ places

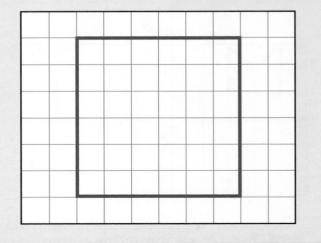

194 one hundred ninety-four CXCIV 97 + 97

Name _____ Date/Time _____

Chapter 9
Lesson 5

Lines of Symmetry

NCTM Standards 2, 3, 6, 7, 8, 9, 10

Look at each picture. Can you divide it into 2 matching parts? If so, draw all the lines of symmetry. If not, write *no*.

1.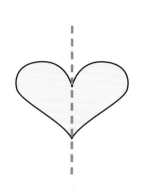

2.

3.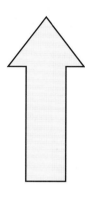

4.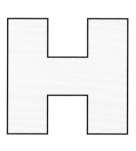

5.

6.

7.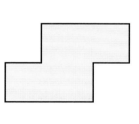

8.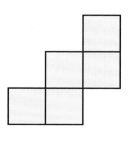

9.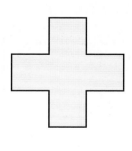

NOTE: Your child is learning about symmetry. Together, look for objects at home that have two identical matching parts when folded on a line.

39   CXCV   one hundred ninety-five   195

Draw the other half of each picture.

10.

11.

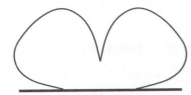

12.

13.

Challenge
Draw a picture for each line of symmetry.

14.

15.

196 one hundred ninety-six   CXCVI   98 + 98

Chapter 9
Lesson 6: Cutting Polygons Apart
NCTM Standards 2, 3, 6, 7, 8, 9, 10

What new figures do you get if you cut along the lines?

1.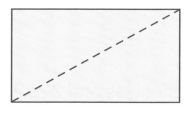

   triangles
   _____

2.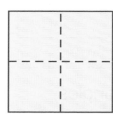

   _____

3.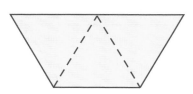

   _____

4.

   _____

5. Cut the rectangle into 2 congruent pieces. Show 2 different ways.

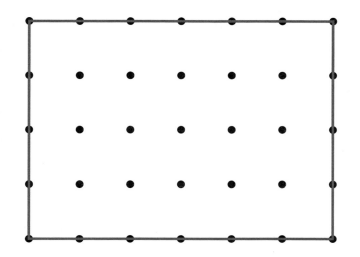 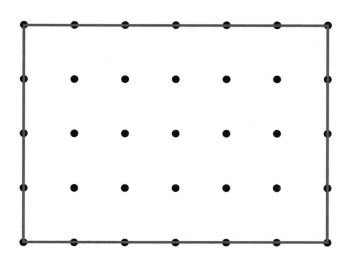

NOTE: Your child is learning to cut polygons to make other polygons. Together, try to cut a sheet of paper into 2 triangles.

100 + 90 + 7   CXCVII   one hundred ninety-seven   197

Draw a line or lines to show the new figures.

6. Make 2 rectangles.

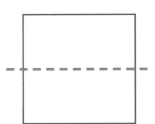

7. Make 2 triangles.

8. Make 4 triangles.

9. Make 2 triangles and 2 rectangles.

10. Mary cut the hexagon from vertex to vertex to make 2 pieces. How many sides would each piece have?

    _____ sides   _____ sides

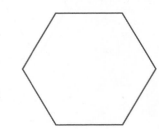

Problem Solving

11. Raffi drew 4 lines that cut her rectangle into only triangles. How many triangles did she make? Draw 2 different ways to do it.

_____ triangles

_____ triangles

198 one hundred ninety-eight CXCVIII 99 + 99

Chapter 9 Lesson 7: Measuring Area

NCTM Standards 1, 2, 3, 4, 6, 7, 8, 9, 10

What is the area? Each ☐ is 1 square unit.

1.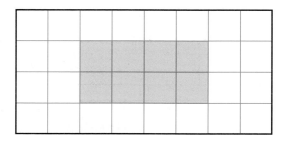

   8 square units

2.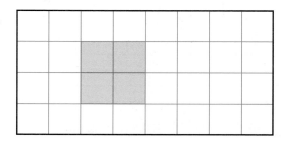

   _____ square units

3.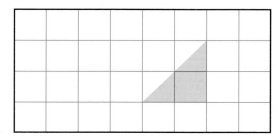

   _____ square units

4.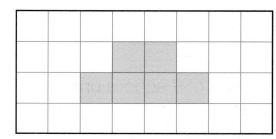

   _____ square units

5.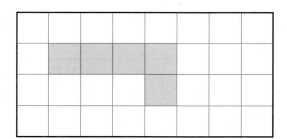

   _____ square units

6.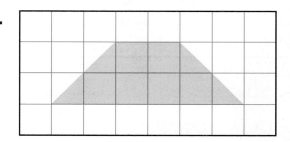

   _____ square units

NOTE: Your child is learning to find the area of a polygon by counting the square units inside the figure.

Draw each figure.

7. a polygon of 5 square units

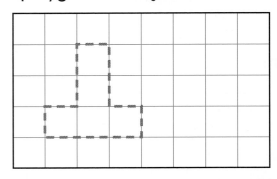

8. a polygon of 6 square units

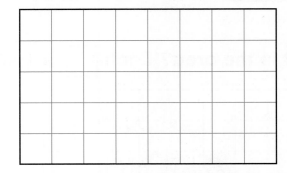

9. a polygon of 3 square units

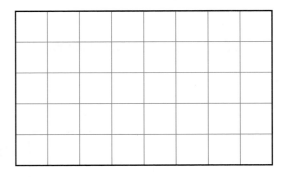

10. a polygon of 8 square units

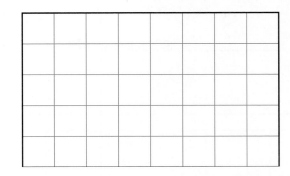

11. a polygon of 9 square units with 8 sides

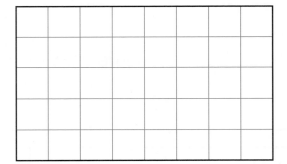

12. a polygon of 5 square units with 8 sides

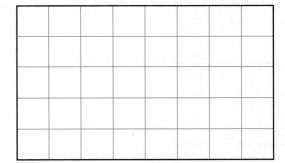

Challenge

13. Draw a polygon with an area of $2\frac{1}{2}$ square units. Explain how to count the area of your polygon.

_____

_____

Chapter 9
Lesson 8: Recording Paths

NCTM Standards 2, 4, 6, 7, 8, 9, 10

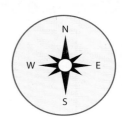

What are the shortest paths from dot to dot? Find all possible ways.

1. from **A** to **B**

   A •—• B

   "Go 1 block East."

   _____

2. from **C** to **D**

   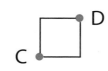

   _____

   _____

3. from **E** to **F**

   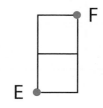

   _____

   _____

   _____

4. from **G** to **H**

   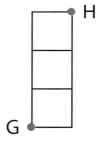

   _____

   _____

   _____

   _____

5. from **I** to **J**

   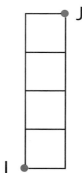

   _____

   _____

   _____

   _____

   _____

6. Complete the table.

Path	Number of Ways
A to B	
C to D	
E to F	
G to H	
I to J	

NOTE: Your child is learning to describe paths on simple maps using notation such as N, S, E, and W to represent the directions North, South, East, and West.

67 + 67 + 67   **CCI**   two hundred one   **201**

What is a shortest path from place to place?

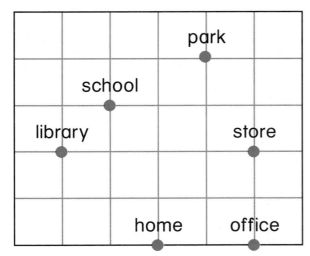
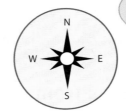

N stands for North.

7. from home to the store ____ ____ ____ ____ ____

8. from the store to school ____ ____ ____ ____ ____

9. from the park to home ____ ____ ____ ____ ____

10. Choose your own starting and ending places.
    Tell what path you will take.

    from _____ to _____ ____ ____ ____ ____ ____

Problem Solving

11. Tina goes to the office and the store. She starts and ends at home. If Tina takes a shortest path, how many blocks does she walk in all? Explain.

    _____ blocks

    _____
    _____
    _____
    _____

Chapter 9 Lesson 9

Directions from Here to There

NCTM Standards 2, 3, 6, 7, 8, 9, 10

Name _____ Date/Time _____

What are the shortest paths from place to place?
Write the shorthand that describes them all.

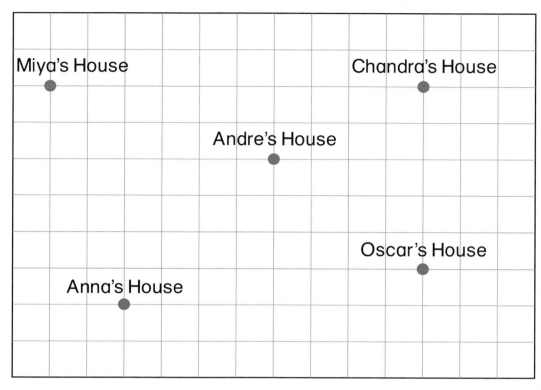
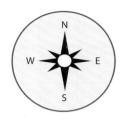

1. from Anna's house to Miya's house   6N 2W _____

2. from Chandra's house to Andre's house   _____

3. from Andre's house to Oscar's house   _____

4. from Anna's house to Andre's house   _____

5. Draw a dot anywhere on the map to show a new house. What is the shortest path from Oscar's house to the new house? _____

NOTE: Your child is learning to describe relative position on a map using the directions north, south, east, and west.

100 + 90 + 13   CCIII   two hundred three   **203**

Use the clues to mark each place on the map.

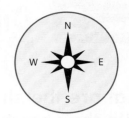
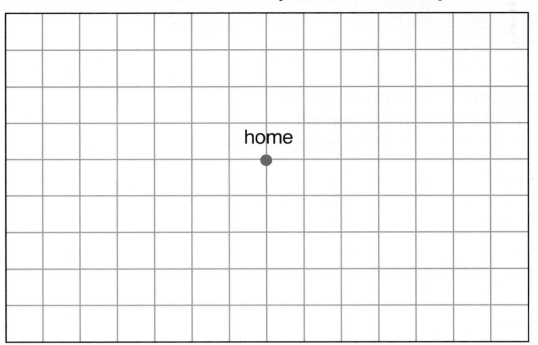

6. **the store**

   The shortest paths from home to the store are 4W 2S.

7. **the school**

   The shortest paths from home to the school are 5E 1N.

8. **the park**

   The shortest paths from the school to the park are 5S 8W.

9. **the library**

   The shortest paths from home to the library are 2S 3E.

10. Compare the shortest paths from home to the store and then from the store to home. What can you say about the lengths of the 2 paths?

    _____

    _____

    _____

Challenge

11. The bank is 4 blocks away from the store. It is 4 blocks away from home. It is 5 blocks away from the park. Mark the bank on the map.

Chapter 9
Lesson 10

Problem Solving Strategy
Draw a Picture

NCTM Standards 1, 2, 3, 6, 7, 8, 9, 10

Understand
Plan
Solve
Check

1. Eve has 3 congruent square tiles. She wants to build a polygon with just 1 line of symmetry. What polygon can she make?

   _____

   _____

   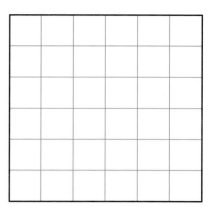

2. Bob wants to cut a rectangle into 2 congruent pieces. What figures can he make?

3. Lisa has 6 game cards. Half of the cards have hearts on them. The rest of the cards have flowers on them. How many of the cards have flowers on them?

   _____ cards

NOTE: Your child is exploring different ways to solve problems. Sometimes drawing a picture is an efficient way to solve a problem.

CCV two hundred five **205**

Problem Solving Test Prep

1. Marie gives $\frac{1}{4}$ of her cookies to Matt. She keeps the 6 cookies that are left. How many cookies did Marie have to start?

   Ⓐ 4 cookies
   Ⓑ 8 cookies
   Ⓒ 12 cookies
   Ⓓ 16 cookies

2. The chef cooks 5 pancakes every 2 minutes. How many pancakes would he make after 10 minutes?

Minutes	2	4	6	8	10
Pancakes	5				

   Ⓐ 9 pancakes
   Ⓑ 10 pancakes
   Ⓒ 20 pancakes
   Ⓓ 25 pancakes

Show What You Know

3. Fran has 28¢. What is the smallest number of coins she could have?

   _____ coins

   Explain how you know that this is the smallest number.

   _____
   _____
   _____
   _____

4. Mike can fit about 25 beans in his hand. He puts the beans in a cup. About how many beans might fill the cup?

   about _____ beans
   Explain how you made your estimate.

   _____
   _____
   _____

Name _____   Date/Time _____

Chapter 9 Review/Assessment
NCTM Standards 2, 3, 4, 6, 7, 8, 9, 10

1. Circle the figures that are polygons. Lesson 1

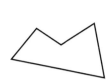

2. Are the figures similar? Write **yes** or **no**. Lesson 3

_____

3. Circle the figure congruent to the blue triangle. Lesson 2

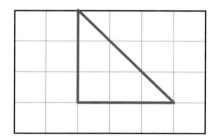
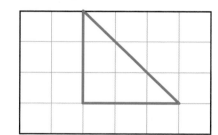
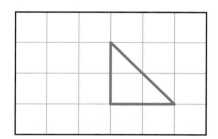

Draw the reflection of each figure. Use a mirror to help you. Lesson 4

4.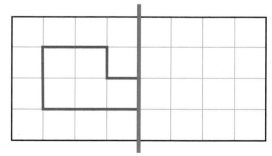

5.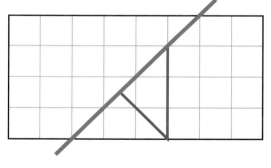

Can you divide each picture into 2 matching parts? If so, draw as many lines of symmetry as you can. If not, write *no*. Lesson 5

6.

7.

69 + 69 + 69   CCVII   two hundred seven   207

What new figures do you get if you cut along the lines? Lesson 6

8.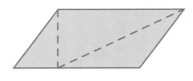

_____

9.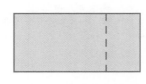

_____

What is the area? Each ☐ is 1 square unit. Lesson 7

10.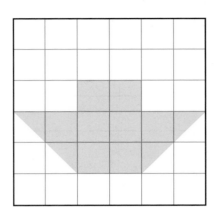

_____ square units

11.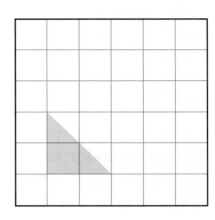

_____ square units

What are the shortest paths from L to M? Write the shorthand. Lessons 8, 9

12. 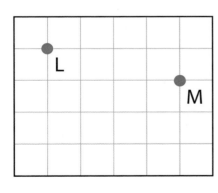 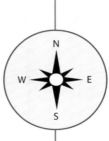 13.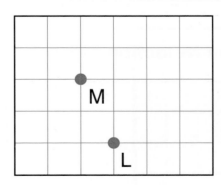

_____ _____

Problem Solving Lesson 10

14. Kelly wants to cut a square into 3 congruent pieces. What figures can she make? Draw a picture.

Chapter 10 Adding and Subtracting Larger Numbers
Double Your Number

You need
- index cards

STEP 1 Reading and Writing Numbers

Pick a card from the pile. Read the numbers on both sides of the card aloud. Write them here.

_____   _____

STEP 2 Comparing

How are both numbers alike? _____
_____

How are they different? _____
_____
_____
_____

STEP 3 Doubling the Numbers

Try to double each number. Show what you do.

_____
_____
_____

School-Home Connection

Dear Family,

Today we started Chapter 10 of *Think Math!* In this chapter, I will develop my ability to add and subtract bigger numbers. There are NOTES on the Lesson Activity Book pages to explain what I am learning every day.

Here are some activities for us to do together at home. These activities will help me understand addition and subtraction.

Love,

_____

Family Fun

Counting Coins

Work with your child to play this game. Your child will play a similar game in Lesson 2.

- You will need a recording sheet, like the one shown below, and a cup with at least 5 pennies, 5 nickels, 5 dimes, and 5 quarters.

Q	D	N	P	Total Value	Fewest Coins?
1	2	1	1	51¢	no

- You and your child take turns picking any 5 coins from the cup and recording in the table.
- Is your amount shown with the fewest coins? If not, then trade to show the fewest coins. Return the coins to the cup each time. Shake the cup before each turn.
- Play until you and your child each take 5 turns.

Shopping for Bargains

Work with your child to identify money amounts in dollar and cents notation.

- Look through store flyers listing prices for items in dollar and cents notation.

- Have children look for patterns in the prices. For example, many of the prices have 99 cents, such as $5.99 or $499.99.

- Together, make up stories about going to the store to buy one of the items.

Chapter 10
Lesson 1: Making Sums of 100

NCTM Standards 1, 2, 6, 7, 8, 9, 10

What number is missing?

1.
```
    25
 + [75]
   ───
   100
```

2.
```
   [ ]
 + 55
   ───
   100
```

3.
```
    23
 + [ ]
   ───
   100
```

4.
```
    81
 + [ ]
   ───
   100
```

5. $92 + \boxed{} = 100$

6. $14 + \boxed{} = 100$

7. $\boxed{} + 48 = 100$

8. $\boxed{} + 66 = 100$

9. Make your own.

$\boxed{} + \boxed{} = 100$

NOTE: Your child is learning to find number pairs with a sum of 100. Ask your child to find the number that when added to 32 makes a sum of 100.

100 + 100 + 11   **CCXI**   two hundred eleven   **211**

10. Which pairs make 100? Circle them as fast as you can.

Sums of 100 Search

36 64	68 43	60 50	51 49	30 75	80 20	12 93	51 64	27 73
60 40	65 10	28 72	85 15	60 30	74 54	29 80	32 68	50 50
76 34	90 10	45 55	37 75	17 83	70 20	62 38	93 16	21 79

11. Pick one of the numbers pairs you just circled. How do you know that it has a sum of 100?

____

____

____

____

Challenge

12. Complete the addition sentence. Then write two subtraction sentences related to it.

22 + ☐ = 100

☐ − ☐ = ☐

☐ − ☐ = ☐

Chapter 10
Lesson 2: Adding with Coins

NCTM Standards 1, 2, 6, 7, 8, 9, 10

What is the value for each collection of coins?

1.

   35 ¢

2.

   _____ ¢

3.

   _____ ¢

4.

   _____ ¢

Show each amount of money with the fewest coins.

5. 35¢

6. 62¢

7. 59¢

8. 40¢

NOTE: Your child is learning to find the value of a collection of coins and to use the fewest coins for that value.

71 + 71 + 71   CCXIII   two hundred thirteen   213

What is missing? Complete the table.

	Number of Coins	Q	D	N	P	Total Value	Is it the fewest coins? If not, draw a picture.
9.	5	1	0	3	1	41¢	Q N D P
10.	8	0	5		3		
11.						35¢	yes
12.	9			3	5	45¢	
13.		2			5	60¢	
14.		0	2	1	4		

Problem Solving

15. Kate has some coins worth 51¢. She could NOT have the fewest coins for that amount. What coins might Kate have? Draw a picture.

Chapter 10 Lesson 3

Patterns in Money

NCTM Standards 1, 2, 6, 9, 10

Name _____ Date/Time _____

At the school store, erasers cost 5¢ and rulers cost 7¢.
What is missing in each price list?

1. **Price List for Erasers**

Number of Erasers	1	2		4		6
Total Cost	5¢		15¢			

2. **Price List for Rulers**

Number of Rulers	1		3	4	
Total Cost	7¢		21¢		

Solve each problem.

3. Billy buys 3 erasers. How much does that cost?

_____ ¢

He gives the clerk 25¢. How much change does he get?

_____ ¢

4. Billy buys 2 erasers and 1 ruler. How much does that cost? _____ ¢

What coins can he use to pay the exact amount? _____

5. Billy gives the clerk 25¢. He buys as many erasers as he can. How many erasers is that?

_____ erasers

How much change does he get?

_____ ¢

6. Billy has 25¢. He buys as many rulers as he can. How many rulers can he get?

_____ rulers

How much does that cost?

_____ ¢

NOTE: Your child is learning to create and extend money patterns to solve problems.

At the school store, pencils cost 8¢ and markers cost 10¢. What is missing in each price list?

7.
Price List for Pencils

Number of Pencils	1	2		4	
Total Cost	8¢		24¢		

8.
Price List for Markers

Number of Markers	1		3	4	
Total Cost	10¢				50¢

Complete the table.

	Billy Has	Billy Buys	Total Cost	Change
9.	50¢	1 pencil, 3 markers		
10.	50¢	2 pencils, 3 markers		
11.	50¢	4 pencils, ___ marker	42¢	
12.	50¢	___ pencils, ___ markers		14¢

Problem Solving

13. Billy wants to spend exactly 50¢. He must buy at least one pencil and one marker. What might he buy?

_____

Chapter 10 Lesson 4

Place Value in Money

NCTM Standards 1, 2, 6, 8, 9, 10

100¢

$1.00

100 cents equals 1 dollar.

What is missing?

	Number of Pennies	Cents	Dollars and Cents
1.	320	320¢	$3.20
2.	86		$0.86
3.	173		
4.		298¢	
5.			$5.04
6.	439		
7.		75¢	

What is each amount in dollars and cents?

8.

_____

9.

_____

 NOTE: Your child is learning to write amounts of money in dollar and cents notation. Together, look through supermarket flyers for prices written with a dollar sign and a decimal point.

100 + 110 + 7 **CCXVII** two hundred seventeen **217**

How can you show each money amount with the fewest dollar bills and coins?

10. $2.36

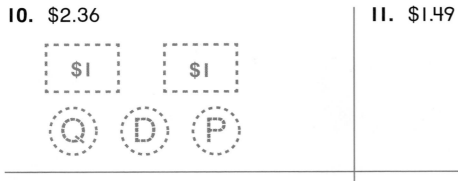

11. $1.49

12. $0.82

13. $3.15

 14. Is $2.28 closer to $2.00 or $3.00? Tell how you know.

___

Challenge
Write each missing amount.

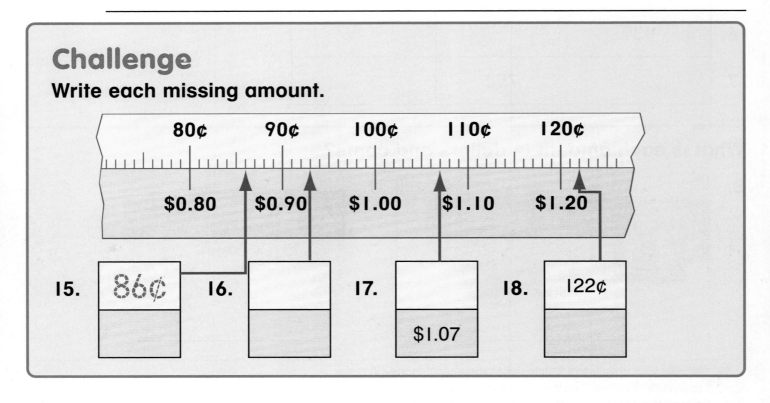

15. 86¢   16.   17.   18. 122¢

$1.07

Chapter 10 Lesson 5

Computing with Money

NCTM Standards 1, 2, 6, 8, 9, 10

What is missing?

1.

10¢	+	4¢	=	14¢
$0.10		$0.04		$

2.

98¢	+	¢	=	
		$0.20		

3.

	+	8¢	=	
$2.53				

4.

	+	124¢	=	
$1.42				

5. $2.00 + $4.00 = $ _6.00_

6. $3.50 + $2.30 = $ ____.____

7. $5.00 + 45¢ = $ ____.____

8. $1.20 + 64¢ = $ ____.____

9. 9¢ + $3.82 = $ ____.____

10. 75¢ + $5.50 = $ ____.____

NOTE: Your child is learning to add and subtract money. Give your child a bill and some coins to make up a problem.

73 + 73 + 73   **CCXIX**   two hundred nineteen   **219**

Use the menu to solve the problems.

11. Sam buys a salad. He pays $5.00. How much is his change?

    $ __1.55__

Menu	
Salad	$3.45
Turkey sandwich ...	$4.70
Milk	85¢
Apple	60¢

12. Sara buys a turkey sandwich. She pays $5.00. How much is her change?

    _____

13. Lee buys an apple. He pays $1.00. How much is his change?

    _____

14. Use the menu to make up your own problem.

    _____

    _____

    _____

Challenge

15. What is missing? Complete the addition table.

+	5¢	25¢	8¢	$1.00	84¢	$
7¢	12¢	32¢		107¢		148¢
	$0.12	$				

220 two hundred twenty   CCXX   44

Chapter 10
Lesson 6: Adding Two-Digit Numbers

NCTM Standards 1, 2, 6, 7, 8, 9, 10

What is missing?

1.

50	7	57
	6	36
	13	

5	7

+ 3	6

9	

2.

	4	14
	3	83

1	4

+ 8	3

3.

		62
		29

6	2

+ 2	9

4.

		28
		65

2	8

+ 6	5

 NOTE: Your child is learning to add numbers using a Cross Number Puzzle. Ask your child to explain how to find the sum of 42 + 39.

100 + 110 + 11   **CCXXI**   two hundred twenty-one   **221**

What is missing?

5.
```
    7  8
 +  3  1
   ___
```
→ (grid showing 78, 31)

6.
```
    5  9
 +  5  7
   ___
```
→ (grid)

7. What is the sum? Tell how you added these numbers.

$23 + 14 = $ _____

_____

_____

_____

_____

Problem Solving

8. There are 19 boys and 17 girls in class. How many children are in class?

_____ children

Show your work.

Chapter 10 Lesson 7: Subtracting Two-Digit Numbers

NCTM Standards 1, 2, 6, 7, 8, 9, 10

What is missing?

1.

10	50	60
	20	26
	30	

⟹

```
   6 0
 - 2 6
 ─────
   3
```

2.

	11	71
10		12

⟹

```
   7 1
 - 1 2
 ─────
```

3.

```
   4 1
 - 1 4
 ─────
```

⟹

30		41
	4	14

4.

```
   6 2
 - 2 6
 ─────
```

⟹

		62
		26

 NOTE: Your child is learning how to subtract numbers using a Cross Number Puzzle. Ask your child to explain how to find the difference for 57 − 23.

200 + 10 + 13   CCXXIII   two hundred twenty-three   **223**

5. What is the difference? Tell how you subtracted these numbers.   58 − 26 = _____

_____

_____

_____

_____

_____

Write a number sentence to describe each problem. Then find the answers.

6. Lisa scored 19 points. Beth scored 23 points. How many points did both girls score?

    _____

    _____ points

7. There were 50 balloons at the store. Manny bought 34 of the balloons. How many balloons were still at the store?

    _____

    _____ balloons

Challenge

8. Mia buys a pencil for 25¢ and an eraser for 35¢. How much change does she get from $1.00?

    _____ ¢

    Show your work.

224 two hundred twenty-four   CCXXIV   112 + 112

Name _____ Date/Time _____

Chapter 10
Lesson 8
Exploring Expanded Notation
NCTM Standards 1, 2, 6, 7, 8, 9, 10

What is missing?

1. 62 = [6 0] + [____]

2. 83 = [____ 0] + [____]

3. 408 = [____ 00] + [____]

4. 375 = [____ 00] + [____ 0] + [____]

5. 981 = [____ 00] + [____ 0] + [____]

6.

400		8	438
	20	0	120
		8	

→

	4	3	8
+	1	2	0

7.

		4	384
500	0		507

→

	3	8	4
+	5	0	7

NOTE: Your child is learning to add and subtract three-digit numbers using expanded notation.

 CCXXV two hundred twenty-five 225

What is each sum or difference?

8.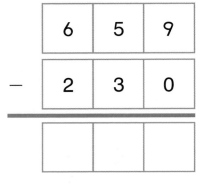

```
    7 1 8
 +  2 0 8
 _____
```

9.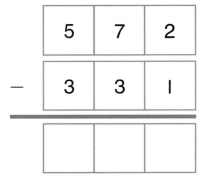

```
    6 5 9
 -  2 3 0
 _____
```

10.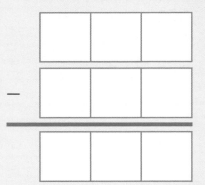

```
    5 7 2
 -  3 3 1
 _____
```

Challenge

11. What subtraction does the puzzle show?

	200		5	
		60		
		80		180

Mental Math with Three-Digit Numbers

Chapter 10, Lesson 9

NCTM Standards 1, 6, 7, 8, 9

Add 138 in pieces to each number! Do all the work in your head. Write only the answers.

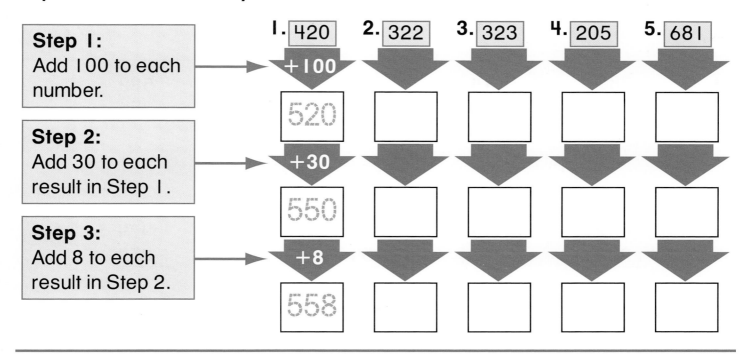

Step 1: Add 100 to each number.

Step 2: Add 30 to each result in Step 1.

Step 3: Add 8 to each result in Step 2.

1. 420 → +100 → 520 → +30 → 550 → +8 → 558
2. 322
3. 323
4. 205
5. 681

Subtract 254 from any number! Do all the work in your head. Write only the answers.

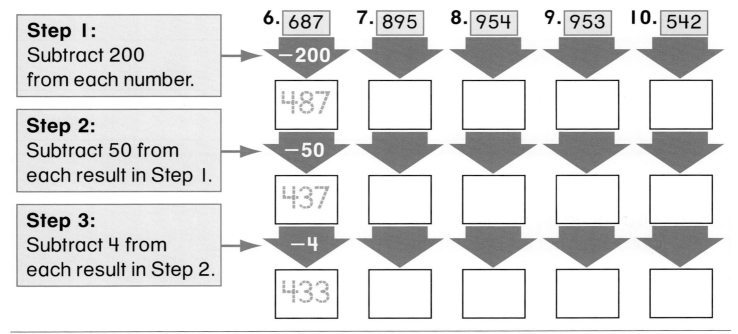

Step 1: Subtract 200 from each number.

Step 2: Subtract 50 from each result in Step 1.

Step 3: Subtract 4 from each result in Step 2.

6. 687 → −200 → 487 → −50 → 437 → −4 → 433
7. 895
8. 954
9. 953
10. 542

NOTE: Your child is learning to add and subtract numbers using mental math. Ask your child to tell how to add 245 + 160 without a pencil and paper.

100 + 110 + 17   CCXXVII   two hundred twenty-seven   **227**

Add or subtract.

Think through all of the calculations in your head.

11.	214 +138	12.	611 +138	13.	180 +138
14.	782 −254	15.	682 −254	16.	582 −254
17.	452 +238	18.	806 −154	19.	474 −354

 20. Write an addition or subtraction story using a problem on this page.

_____

_____

_____

Problem Solving

21. Lisa added 200 to her number and got 541. What is Lisa's number? _____

22. Conor added 50 to his number and got 684. What is Conor's number? _____

Chapter 10
Lesson 10
Adding Two- and Three-Digit Numbers

NCTM Standards 1, 2, 6, 8, 9, 10

What is each sum?

1.
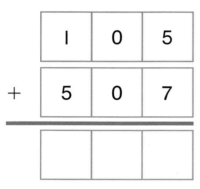

	2	1	4
+	7	3	8
			2

200	10	4	214
			738

2.

	1	0	5
+	5	0	7

3.	123 +321	4.	241 +517	5.	146 + 32	6.	319 +435
7.	428 +537	8.	842 + 73	9.	408 +337	10.	318 +152

NOTE: Your child is practicing addition. Ask your child to add this page number to the next page number and tell you the sum.

100 + 120 + 9   CCXXIX   two hundred twenty-nine   229

11. Which two numbers in the box have a sum of about 400?

127	976
325	438
509	42

_____ and _____

12. Which two numbers in the box have a sum of about 900?

_____ and _____

Write a number sentence to solve each problem.

13. There are 127 children in the first grade and 146 children in the second grade. How many children are in the two grades?

_____ + _____ = _____   _____ children

14. Shaun read a book with 213 pages. He read another book with 254 pages. How many pages did Shaun read in both books?

_____ + _____ = _____   _____ pages

Challenge
What is each sum?

15.
```
  132
  127
+ 203
─────
```

16.
```
  215
  120
+  64
─────
```

Name _____ Date/Time _____

Chapter 10
Lesson 11

Subtracting Two- and Three-Digit Numbers

NCTM Standards 1, 2, 6, 9, 10

What is each difference?

1.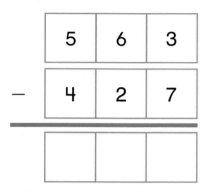

2.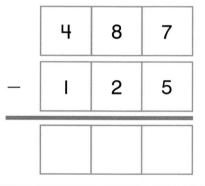

3.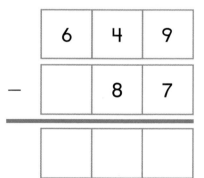

4.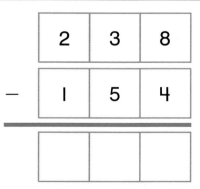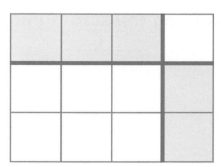

 NOTE: Your child is practicing subtraction. Ask your child to subtract this page number from 480.

77 + 77 + 77   CCXXXI   two hundred thirty-one **231**

What is each difference?

5. 789 −310 **479**	6. 645 −222	7. 593 −348	8. 375 − 59
9. 428 −163	10. 247 − 83	11. 708 −276	12. 417 −352

Write a number sentence to solve each problem.

13. There are 863 children at Parkside Elementary School. There are 124 children in the second grade. How many children are NOT in the second grade?

    _____ − _____ = _____   _____ children

14. Loni is reading a book with 257 pages. She has already read 36 pages. How many pages does Loni have left to read?

    _____ − _____ = _____   _____ pages

Challenge
What is each missing number?

15. ☐
    − 352
    ─────
     176

16.   685
    − ☐
    ─────
     243

232   two hundred thirty-two   CCXXXII   116 + 116

Chapter 10
Lesson 12 — Practice Adding and Subtracting
NCTM Standards 1, 2, 6, 8, 9, 10

What is each sum or difference?

1. 573 +321 *894*	2. 605 +143	3. 784 −412	4. 496 − 73
5. 231 + 45	6. 591 −245	7. 427 +316	8. 829 −356
9. 362 +418	10. 605 −143	11. 372 + 45	12. 427 −356

13. Use each of the digits 3, 2, and 5 once.

   Make the biggest number. _____

   Make the smallest number. _____

   Find the sum. 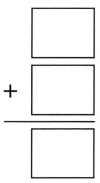   Find the difference.

NOTE: Your child is practicing addition and subtraction. Read your child's answers to Problems 14 and 15.

100 + 120 + 13   **CCXXXIII**   two hundred thirty-three **233**

14. What is the sum? Describe how you added the numbers.

    358
    +426
    _____

    _____

    _____

    _____

    _____

15. What is the difference? Describe how you subtracted the numbers.

    643
    −217
    _____

    _____

    _____

    _____

    _____

Problem Solving

Would you add or subtract to solve the problem? Circle the correct operation.

16. Harry plays soccer for 45 minutes on Sunday and for 90 minutes on Monday. How long does he play soccer in all?

    add   subtract

17. Lorri buys a notebook for $1.99. She pays the clerk with a $5.00 bill. How much change does she get?

    add   subtract

Chapter 10
Lesson 13

Problem Solving Strategy
Solve a Simpler Problem

NCTM Standards 1, 2, 3, 4, 6, 7, 8, 9, 10

Understand
Plan
Solve
Check

1. Maria bought 2 erasers. Each eraser costs 48¢. How much did Maria spend for the erasers?

   _____ ¢

   How did you find the answer?

   _____

   _____

   _____

2. Gabe goes to the library every 4 days. He went on the 4th day of the year. Will he go to the library on the 365th day of the year?

   _____

   How did you find the answer?

   _____

   _____

   _____

3. Joe delivers 60 newspapers every day. On Sundays he delivers an extra 32 papers. How many newspapers does Joe deliver in one week?

   _____ newspapers

   How did you find the answer?

   _____

   _____

   _____

NOTE: Your child is exploring different ways to solve problems. Sometimes solving a simpler problem is an efficient way to solve a problem.

47

CCXXXV   two hundred thirty-five   235

Problem Solving Test Prep

1. The first three doors in a hallway are numbered 143, 145, and 147. What is the number of the fifth door?

   Ⓐ 141

   Ⓑ 149

   Ⓒ 150

   Ⓓ 151

2. The movie ended at 8:40. It was 1 hour and 40 minutes long. What time did the movie start?

   Ⓐ 6:00

   Ⓑ 7:00

   Ⓒ 7:20

   Ⓓ 10:20

Show What You Know

3. Carl wants to cut a square into 4 congruent pieces. Name two figures he could make.

   _____

   _____

   _____

   Explain how you found the answer.

   _____

   _____

   _____

4. If 2 people share some cookies, they each get an odd number of cookies. If 3 people share the cookies, they each get an even number of cookies. How many cookies could there be?

   _____ cookies

   Explain how you know.

   _____

   _____

   _____

   _____

   _____

   _____

   _____

Chapter 10 Review/Assessment

NCTM Standards 1, 2, 6, 8, 9, 10

What number is missing? Lesson 1

1.
```
   85
+ ☐
-----
  100
```

2.
```
   51
+ ☐
-----
  100
```

What is the value of each collection of coins? Lesson 2

3.

_____ ¢

4.

_____ ¢

5. Each sticker costs 6¢. What is missing in the price list? Lesson 3

Price List for Stickers

Number of Stickers	1	2		4		6
Total Cost	6¢		18¢			

6. What is the amount in dollars and cents? Lesson 4

_____

7. What is the sum? Lesson 5

$1.25 + $2.30 = $ _____

79 + 79 + 79   CCXXXVII   two hundred thirty-seven   237

What is missing? Lessons 6, 7, 8

8.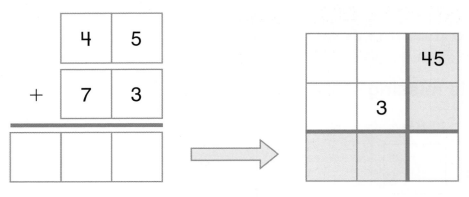

9.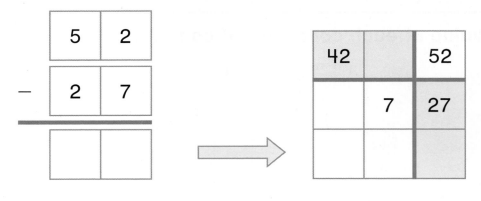

10. 461 = [____00] + [____0] + [____]

Find the sum or difference. Lessons 9, 10, 11, 12

11. 239
    + 29
    ───

12. 560
    +317
    ───

13. 756
    −545
    ───

14. 900
    −730
    ───

Problem Solving Lesson 13

15. Sean gives the cashier $8.00. He gets $6.10 in change. How much does he spend?

$_____

Chapter 11

Name _____

Skip-Counting and Equivalent Sets
Making a Quilt

You need
- 4 squares of paper or cloth
- regular or fabric markers

Make a quilt by arranging squares in a pattern.

STEP 1 — Making Rectangles

Arrange your squares to make a rectangle. How many different rectangles can you make with all four squares?   _____ rectangles

Draw the different rectangles.

STEP 2 — Making a Group Quilt

Combine your squares with the others in your group. Make a big quilt in the shape of a rectangle. Draw the quilt.

STEP 3 — Making Quilt Patterns

Decorate the quilt with a pattern. Describe your pattern.

_____

_____

_____

Investigation

School-Home Connection

Dear Family,

Today we started Chapter 11 of *Think Math!* In this chapter, I will explore how to multiply by combining equivalent sets and how to divide by making fair shares. There are NOTES on the Lesson Activity Book pages to explain what I am learning every day.

Here are some activities for us to do together at home. These activities will help me understand multiplication and division.

Love,

_____

Family Fun

How Many?

Work with your child to play this game. Your child will play a similar game in Lesson 2.

- You will need a recording sheet like the one shown below, a number cube, and pennies or other small items like buttons or cereal pieces.

How many items are in each set?	How many sets are there?	How many items are there in all?

- You and your child take turns. For each turn, toss a number cube two times. The first toss shows how many items to put in a set. The second toss shows how many sets to make.
- Find the total number of items.
- Play until you and your child each take 5 turns.

Sharing Cookies

Work with your child to share amounts of cookies fairly.

- You will need 3 plates and a handful of cookies or other small food items.
- Count out any number of cookies into a pile.
- Together, see if the cookies can be shared fairly among 3 people by placing the cookies on the plates.

- Could you share the cookies fairly? Try other amounts of cookies and see which amounts can be shared fairly and which cannot.

Chapter 11 Lesson 1: Looking For Patterns in Jumps

NCTM Standards 1, 2, 6, 8, 9, 10

Skip-count on the number lines. Label your jumps.

1.
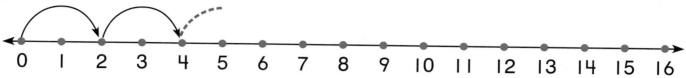

2.
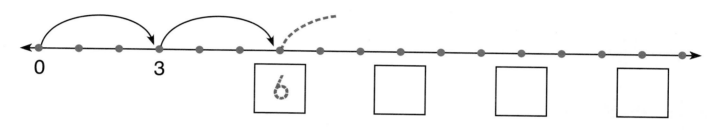

3.
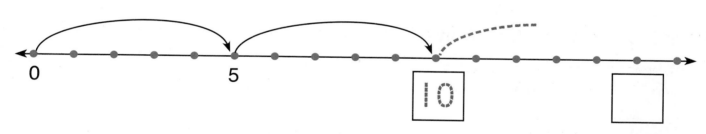

What numbers come next in each pattern?

		1st	2nd	3rd	4th	5th	6th	7th	8th	9th
4.	2↷	2	4	6						
5.	3↷	3	6	9						
6.	5↷	5	10	15						

NOTE: Your child is working more with skip-counting. Together, look for patterns when skip-counting by fives.

100 + 130 + 11   **CCXLI**   two hundred forty-one   **241**

Skip-count on the number lines. Label your jumps.

7.

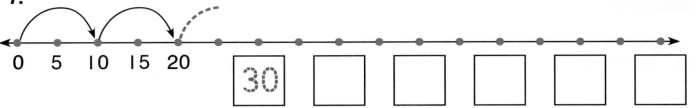

0 5 10 15 20   [30] [ ] [ ] [ ] [ ] [ ]

8.

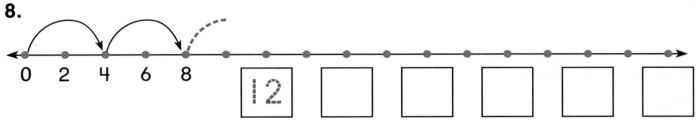

0 2 4 6 8   [12] [ ] [ ] [ ] [ ] [ ]

9.

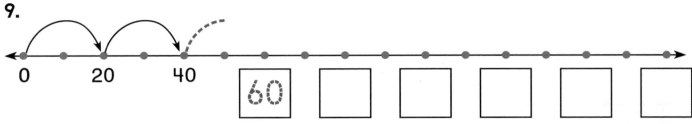

0   20   40   [60] [ ] [ ] [ ] [ ] [ ]

What numbers come next in each pattern?

10.	10	10	20	30				
11.	4	4	8	12				
12.	20	20	40	60				

Problem Solving

13. Mike gets 1 nickel each day. How much money will he have in 7 days? Explain. _____¢

_____

_____

242 two hundred forty-two CCXLII 121 + 121

**Chapter 11
Lesson 2**

Combining Equivalent Sets

NCTM Standards 1, 2, 6, 7, 8, 9, 10

How many are there in all?

1.

   This is 4 sets of 3, or four threes.

   __12__ counters

2.

   _____ counters

3.

   _____ cubes

4.

   _____ cubes

5. Draw equivalent sets. Find how many in all.

NOTE: Your child is learning to combine equivalent sets of objects and find the total.

81 + 81 + 81   **CCXLIII**   two hundred forty-three   **243**

Draw sets of circles. How many are there in all?

6. 3 sets of 6

   _____ in all

7. 4 sets of 5

   _____ in all

8. Make your own.

   _____ sets of _____   _____ in all

Problem Solving

9. Luke buys 3 bags of carrots with 5 carrots in each bag. Molly buys 2 bags of carrots with 10 carrots in each bag. Who buys more carrots? _____

   Use words, numbers, or pictures to explain.

Chapter 11 Lesson 3

Organizing Equivalent Sets

NCTM Standards 1, 2, 6, 7, 8, 9, 10

There will be 4 children at Lynn's party. Each child will get 9 cookies. How many cookies does Lynn need to make?

1. Put the cookies in rows so they are easier to count. Draw them on the cookie sheet. Write how many.

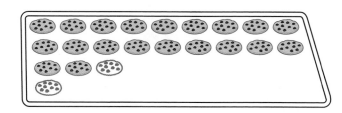

___4___ rows of _____ cookies

_____ cookies in all

2. Now draw the cookies in this grid. Write how many.

___4___ rows and _____ columns

_____ cookies in all

NOTE: Your child is learning to arrange objects in an array to make counting easier.

CCXLV   two hundred forty-five   245

What is missing?

3.

Number of rows	Number in each row	Total
3	4	12

4.

Number of rows	Number in each row	Total
2		

5.

Number of rows	Number of columns	Total
	5	

6.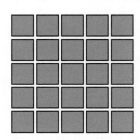

Number of rows	Number of columns	Total

7.

Number of rows	Number of columns	Total
	3	

8.

Number of rows	Number of columns	Total

Problem Solving

9. Bob has 3 rows of 8 chairs. How else could he put all of the chairs in equal rows? Use words, numbers, or pictures to explain.

Chapter 11
Lesson 4

Adding Equivalent Sets

NCTM Standards 1, 2, 5, 6, 7, 8, 9, 10

How many are there in all? Write an addition sentence.

1.

   _4_ + _4_ + _4_ = _12_

2.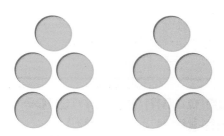

   ___ + ___ = ___

3.

   ___ + ___ + ___ = ___

4.

   _____

5.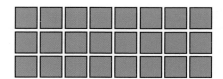

   _____

6.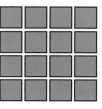

   _____

NOTE: Your child is learning to write addition sentences to add equivalent sets.

100 + 140 + 7   CCXLVII   two hundred forty-seven   **247**

The town keeps track of how many vehicles use their bridge every day.

Kind of Vehicle	Vehicles Using the Bridge Today							
cars	☺	☺	☺	☺	☺	☺	☺	
trucks	☺	☺	☺					
vans	☺	☺	☺					
buses	☺	☺						

Key: Each ☺ stands for 8 vehicles.

Write an addition sentence to find the total for each kind of vehicle.

7. cars _____ _____ cars

8. trucks _____ _____ trucks

9. vans _____ _____ vans

10. buses _____ _____ buses

 11. Write your own question about the pictograph. Show how to solve the problem.

_____

_____

Problem Solving

12. Alex wrote 3 + 3 + 3 + 3 + 3 = 15 for an array. What other number sentence could he write for the same array?

_____

Draw a picture to explain.

248 two hundred forty-eight   CCXLVIII   124 + 124

Chapter 11 Lesson 5: Working with Rectangular Arrays

NCTM Standards 1, 2, 6, 7, 8, 9, 10

What is missing?

1.

Rows	Columns	Squares
2	9	

___2___ × ___9___ = _____

2.

Rows	Columns	Squares

_____ × _____ = _____

3.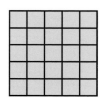

Rows	Columns	Squares

_____ × _____ = _____

4.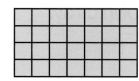

Rows	Columns	Squares

_____ × _____ = _____

5.

Rows	Columns	Squares

_____ × _____ = _____

6.

Rows	Columns	Squares

_____ × _____ = _____

NOTE: Your child is learning to write multiplication sentences for arrays.

83 + 83 + 83   **CCXLIX**   two hundred forty-nine

Write one addition sentence and one multiplication sentence for each array.

7.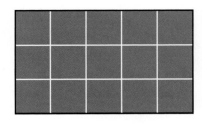

____3____ × ____5____ = _____

5 + 5 + 5 =
_____

8.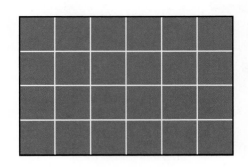

_____ × _____ = _____

_____

9. Make your own array.

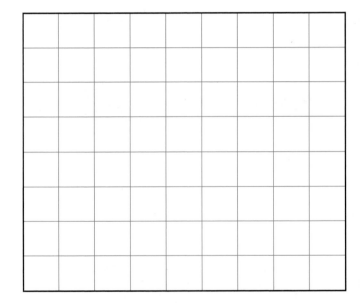

_____ × _____ = _____

_____

Challenge

10. Write two different multiplication sentences for this array.

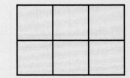

_____ × _____ = _____   _____ × _____ = _____

Chapter 11 Lesson 6

Building Multiples

NCTM Standards 1, 2, 6, 8, 9, 10

Name _____   Date/Time _____

How many are there?

1. How many eggs are in the box?

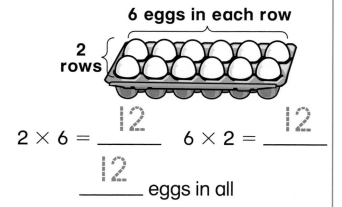

6 eggs in each row
2 rows

2 × 6 = __12__   6 × 2 = __12__

__12__ eggs in all

2. How many cookies are on the tray?

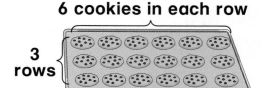

6 cookies in each row
3 rows

3 × 6 = _____   6 × 3 = _____

_____ cookies in all

3. How many legs are on 3 dogs?

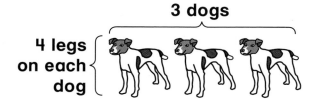

3 dogs
4 legs on each dog

4 × 3 = _____   3 × 4 = _____

_____ legs in all

4. How many flowers are in 2 vases?

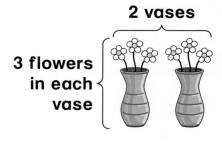

2 vases
3 flowers in each vase

2 × 3 = _____   3 × 2 = _____

_____ flowers in all

5. How many fingers are on 4 hands?

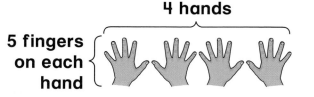

4 hands
5 fingers on each hand

4 × 5 = _____   5 × 4 = _____

_____ fingers in all

6. How many days are in 4 weeks?

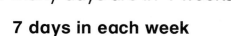
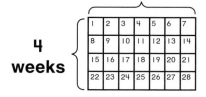

7 days in each week
4 weeks

7 × 4 = _____   4 × 7 = _____

_____ days in all

NOTE: Your child is learning to find multiples of a number. Together find how many shoes are in the house.

100 + 140 + 11   **CCLI**   two hundred fifty-one   **251**

Complete each table.

7. How many wheels are on 9 tricycles?

Number of Tricycles	1	2	3	4	5	6	7	8	9
Number of Wheels	3	6							

8. How many legs are on 9 chairs?

Number of Chairs	1	2	3	4	5	6	7	8	9
Number of Legs	4	8							

9. A spider has 8 legs. How many legs do 9 spiders have? Make a table to find out.

_____ legs

Problem Solving

10. Steve has 6 weeks to finish a project. There are 7 days in a week. How many days does Steve have to finish the project?

_____ days

Chapter 11 Lesson 7

Sharing Between Two Children

NCTM Standards 1, 2, 6, 7, 8, 9, 10

How many does each child get? Draw to share each amount equally between 2 children.

Try it with counters.

1. 16 cookies

 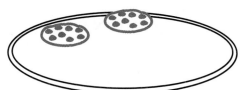

Each child gets __8__ cookies.

2. 10 cookies

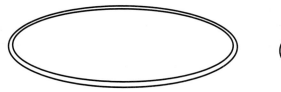 

Each child gets ____ cookies.

3. 22 cookies

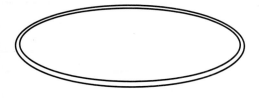 

Each child gets ____ cookies.

4. Make your own. ____ cookies

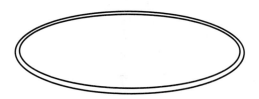 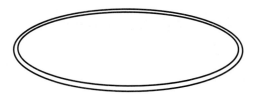

Each child gets ____ cookies.

 NOTE: Your child is learning to divide amounts into two equivalent sets.

200 + 40 + 13   CCLIII   two hundred fifty-three **253**

Share each amount in 2 equivalent sets.
Use a different color for each set.

5. ●●○○○○○○○○○○○○

| 14 balls | ⇒ | 2 shares | ⇒ | _7_ balls each |

6. ♡♡♡♡♡♡♡♡♡♡♡♡♡♡♡♡♡♡♡♡

| 20 hearts | ⇒ | 2 shares | ⇒ | _____ hearts each |

7. ▫▫▫▫▫▫▫▫▫▫▫▫▫▫▫▫▫▫

| 18 blocks | ⇒ | 2 shares | ⇒ | _____ blocks each |

8. Draw any number of objects. Try to make 2 equivalent sets. Explain what you did.

_____

_____

_____

Problem Solving

9. I have 8 of the same coins. I share the coins equally in 2 pockets. How much money might be in each pocket?

   _____ ¢

Show how you solved the problem.

254 two hundred fifty-four CCLIV 127 + 127

Chapter 11 Lesson 8

Sharing Among Three Children

NCTM Standards 1, 2, 6, 7, 8, 9, 10

Name _____ Date/Time _____

How many does each child get? Draw to share each amount equally among 3 children.

1. 12 cookies

Each child gets ____4____ cookies.

2. 21 cookies

Each child gets _____ cookies.

3. 30 cookies

Each child gets _____ cookies.

4. Make your own. _____ cookies

Each child gets _____ cookies.

NOTE: Your child is learning to divide amounts equally into three sets. Together, try to divide a bunch of pennies equally into 3 piles.

CCLV   two hundred fifty-five   **255**

Share each amount in 3 equivalent sets.
Use a different color for each set.

5.

   9 balls ⟹ 3 shares ⟹ 3 balls each

6.

   24 hearts ⟹ 3 shares ⟹ _____ hearts each

7.

   15 blocks ⟹ 3 shares ⟹ _____ blocks each

8. Look back at Problems 5 to 7. Which of these sets can also be divided into 2 equivalent sets? Tell how you know.

   _____

   _____

   _____

Challenge

9. What is missing?

3	6	9			18	21	
1	2	3	4				8

256 two hundred fifty-six   CCLVI   128 + 128

Chapter 11
Lesson 9: How Many Packages?

NCTM Standards 1, 2, 6, 8, 9, 10

How many packages can you fill? Complete each order. Use counters or draw a picture.

The factory can put any number of wheels in a package for special orders.

1. Start with 15 wheels.
   Put 5 in each package.

   Fill __3__ packages.

2. Start with 24 wheels.
   Put 4 in each package.

   Fill _____ packages.

3. Start with 27 wheels.
   Put 3 in each package.

   Fill _____ packages.

4. Start with 48 wheels.
   Put 6 in each package.

   Fill _____ packages.

Make your own.

5. Start with _____ wheels.

   Put _____ in each package.

   Fill _____ packages.

 NOTE: Your child is learning to divide amounts into equivalent sets and find how many sets.

100 + 150 + 7   CCLVII   two hundred fifty-seven   257

How many sets can you make?

6.
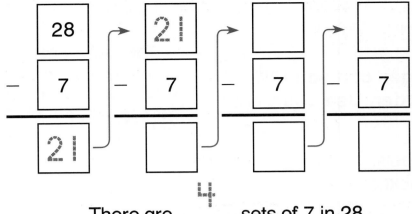

There are ____4____ sets of 7 in 28.

7.
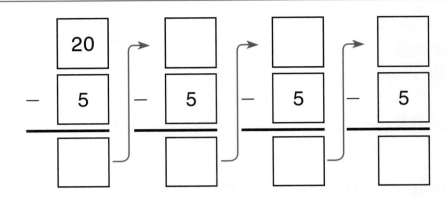

There are _____ sets of 5 in 20.

8.
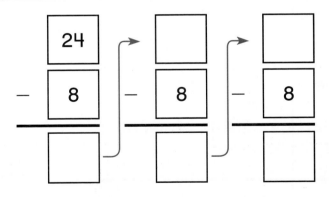

There are _____ sets of 8 in 24.

Problem Solving

9. Larry is packing an order of wheels. He fills 3 packages of 6 wheels each. He has 12 wheels left to pack. How many wheels are in the total order? _____ wheels

Chapter 11 Lesson 10

Problem Solving Strategy
Make a List

NCTM Standards 1, 2, 4, 5, 6, 7, 8, 9, 10

Understand
Plan
Solve
Check

1. Erasers come in packs of 4. Pencils come in packs of 6. I want to buy the same number of erasers and pencils. How many packs of erasers and pencils do I need to buy?

   _____ packs of erasers

   _____ packs of pencils

   1. 4 erasers, 6 pencils
   _____
   _____
   _____
   _____
   _____

2. Paula has 3 shirts. One is red, one is yellow, and one is green. She has two skirts. One is black and one is white. How many different outfits can she make?

   _____ outfits

3. Sid uses a toothpick to make each side of a triangle. He makes 5 triangles. How many toothpicks does he need?

   _____ toothpicks

4. Lisa, Max, and Nate are running in a race. They finish first, second, and third. How many different ways can they finish the race?

   _____ ways

NOTE: Your child is exploring different ways to solve problems. Sometimes making a list is an efficient way to solve a problem.

200 + 50 + 9   **CCLIX**   two hundred fifty-nine   **259**

Problem Solving Test Prep

1. At a bus stop, 3 people get on and 2 people get off. Now there are 26 people on the bus. How many people were on the bus before it stopped?

    Ⓐ 5 people

    Ⓑ 25 people

    Ⓒ 26 people

    Ⓓ 28 people

2. A snail travels 1 foot every 5 minutes. He starts crawling at 6:15. What time will it be when he has traveled 10 feet?

    Ⓐ 6:05

    Ⓑ 6:20

    Ⓒ 6:50

    Ⓓ 7:05

Show What You Know

3. David has 36 chairs. Half of the chairs have pads. How many chairs do not have pads?

    _____ chairs

    Explain how you found the answer.

    _____
    _____
    _____
    _____
    _____

4. Doris has 3 quarters. She wants to buy two notebooks. Each notebook costs 49¢. Does she have enough to buy both notebooks?

    _____

    Explain how you know.

    _____
    _____
    _____
    _____
    _____

Chapter 11 Review/Assessment

NCTM Standards 1, 2, 6, 7, 8, 9, 10

1. Skip count on the number line. Label your jumps. Lesson 1

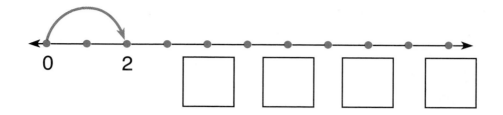

2. How many are there in all? Lesson 2

   _____ counters in all

3. What is missing? Lesson 3

Number of rows	Number in each row	Total
2		

How many are there in all? Write an addition sentence. Lesson 4

4.

   ___ + ___ + ___ = ___

5.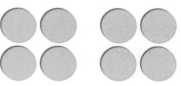

   ___ + ___ = ___

6. Write one addition sentence and one multiplication sentence for the array. Lesson 5

   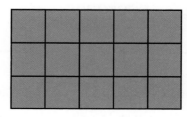

   _____

   ___ × ___ = ___

87 + 87 + 87   **CCLXI**   two hundred sixty-one   261

7. How many in all? Complete the table. Lesson 6

Number of Hands	1	2	3	4	5	6	7	8	9
Number of Fingers	5	10							

8. How many does each child get? Draw to share 12 cookies equally between 2 children. Lesson 7

Each child gets _____ cookies.

9. Share 6 balls equally among 3 children. Lesson 8

6 balls ➡ 3 shares ➡ _____ balls each

10. How many packages can you fill? Complete the order. Use counters or draw a picture. Lesson 9

Start with 16 wheels.
Put 4 in each package.

Fill _____ packages.

Problem Solving Lesson 10

11. Mary uses a toothpick to make each side of a square. She makes 5 squares. How many toothpicks does she need?

_____ toothpicks

Name _____

Chapter 12 Measuring Length
How Far Is It?

You need
- measuring tools such as yardstick, ruler, string, paper clips, and tape measure

Try different ways to measure distance.

STEP 1 Estimating Distance

Stand on one side of your classroom. Look across the room. Estimate the distance to the other side of the room.

It looks about _____ long.

STEP 2 Measuring Distance

Talk with your group about how you could measure the length of the room. Then try it. What did you find?

_____

_____

It measures about _____ long.

STEP 3 Comparing Lengths

Compare lengths with other groups. How were the results alike? How were they different?

_____

_____

_____

_____

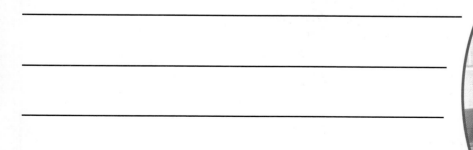

Investigation

School-Home Connection

Dear Family,
Today we started Chapter 12 of *Think Math!* In this chapter, I will learn different ways and different units to measure length. There are NOTES on the Lesson Activity Book pages to explain what I am learning every day.

Here are some activities for us to do together at home. These activities will help me as I learn to measure length.

Love,

_____

Family Fun

What's My Length?

Work with your child to play this game. Your child will play this game in Lesson 4.

- Think of a simple measurement in inches, feet, or yards. For example, you might use 1 foot, 6 inches, or 5 yards. Do not share the measurement with your child. Keep it a secret.

- Have your child ask *yes/no* questions to try and guess your secret measurement. Some possible questions might be:
  - Is it longer than 1 foot?
  - Is it as long as this room?
  - Is it shorter than my hand?

- Have your child continue asking questions until he or she has correctly identified the measurement.

- Switch roles and play again.

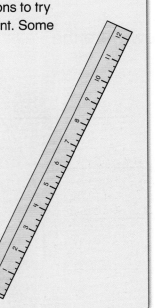

Reading Road Signs

Together, find road signs listing different distances.

- When traveling around town or on a highway, play an informal game to find road signs that give different distances.

- Call them out as you see them. After reading each one, tell whether it is a short distance or a long distance.

- After returning home, you might want to draw some of the signs you saw on your trip.

Chapter 12 Lesson 1

Name _____ Date/Time _____

Measuring Length with Nonstandard Units
NCTM Standards 1, 4, 6, 9, 10

My group is measuring with _____ as the unit.

Draw a line for each length.

1. 1 unit long

   •

2. 2 units long

   •

3. 3 units long

   •

4. 4 units long

   •

5. Make your own line. Measure it and write the length.

   •

   about _____ units long

 NOTE: Your child is learning about nonstandard units. Have your child measure objects around your home using pennies as a unit of measure.

53   CCLXV   two hundred sixty-five   265

How long is each bar? Use your unit to measure.

6.

   The bar is about _____ units long.

7.

   The bar is about _____ units long.

8.

   The bar is about _____ units long.

9. Draw a bar. Find its length with your unit.

   The bar is about _____ units long.

Challenge

10. How long is the path? Use your unit.

    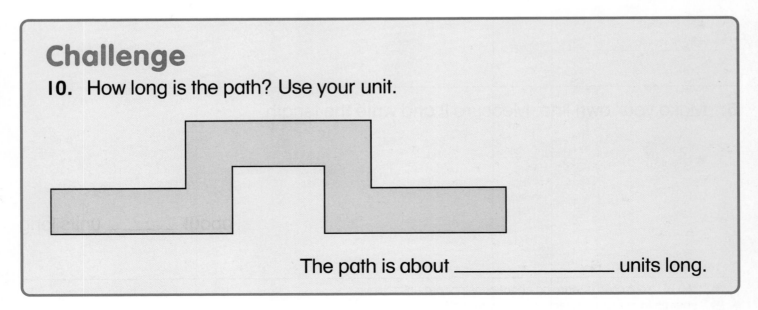

    The path is about _____ units long.

Chapter 12
Lesson 2

Measuring to the Nearest Inch

NCTM Standards 1, 4, 6, 8, 9, 10

How long is the picture of each object? Use a ruler to measure to the nearest inch.

1.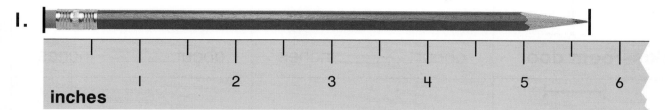

   about ___6___ inches

2.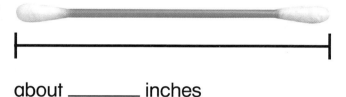

   about _____ inches

   Line up the end of the ruler with the end of the object.

3.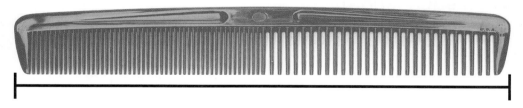

   about _____ inches

4.    about _____ inches

5.    about _____ inch

 NOTE: Your child is learning to measure length to the nearest inch.

89 + 89 + 89   CCLXVII   two hundred sixty-seven   **267**

How long is each real object? Estimate.
Then measure to the nearest inch.

	Object	Estimate	Measurement
6.	classroom door	about _____ inches	about _____ inches
7.	desk	about _____ inches	about _____ inches
8.	math book	about _____ inches	about _____ inches

Draw a picture of something in your classroom that matches each length.

9. about 2 inches

10. about 5 inches

11. about 10 inches

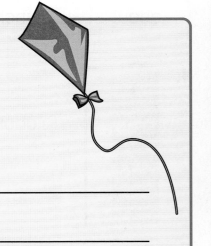

Problem Solving

12. Todd is buying a new kite. He wants one that is more than 10 inches wide. He goes to the store without a ruler. How might he measure the kite?

_____

_____

268 two hundred sixty-eight   CCLXVIII   134 + 134

Chapter 12 Lesson 3
Measuring in Inches, Feet, and Yards
NCTM Standards 1, 4, 6, 8, 9, 10

Which unit would you use to measure each real object? Circle *inches*, *feet*, or *yards*.

1.

   inches
   feet
   (yards)

2.

   inches
   feet
   yards

3.

   inches
   feet
   yards

4.

   inches
   feet
   yards

5.

   inches
   feet
   yards

6.

   inches
   feet
   yards

7.

   inches
   feet
   yards

8.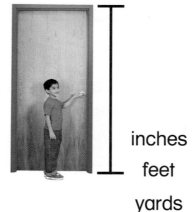

   inches
   feet
   yards

9.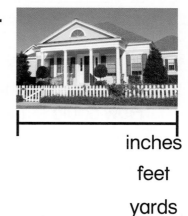

   inches
   feet
   yards

NOTE: Your child is learning about inches, feet, and yards. Ask your child to name an object or distance that would be best to measure in yards.

200 + 60 + 9   **CCLXIX**   two hundred sixty-nine   **269**

Draw a classroom object or distance you would measure with each unit. Find the length.

	Unit	Object	Length to the Nearest Unit
10.	inches		_____ inches
11.	feet		_____ feet
12.	yards		_____ yards

Write *inch*, *foot*, or *yard* for each benchmark object.

13.

about 1 _____

14.

about 1 _____

15.

a little more than 1 _____

Problem Solving

16. Jan says that she measured her book with string. How could string help her find the width?

_____

_____

Chapter 12
Lesson 4: Relating Inches, Feet, and Yards

NCTM Standards 1, 2, 4, 6, 8, 9, 10

Which is the best unit to measure the real object?
Write *inches, feet,* or *yards*.

1.

   about 4 _____ long

2.

   about 6 _____ long

3.

   about 5 _____ long

4.

   about 5 _____ long

5.

   about 6 _____ tall

 NOTE: Your child is learning how inches, feet, and yards are related. Ask your child to tell how many inches are in 1 foot and how many feet are in 1 yard.

200 + 60 + 11   CCLXXI   two hundred seventy-one   271

Complete each table.

6.

Number of Yards	1	2	3		5	6	
Number of Feet	3			12			21

7.

Number of Feet	1	2	3	4	5		7
Number of Inches	12	24				72	

Use the tables to complete the problems.
Then write your own problems.

8. The chalkboard is 2 yards long. How many feet is that?

_____ feet

9. The baseball bat is 3 feet long. How many inches is that?

_____ inches

10. Write a problem about yards and feet.

_____

_____

_____

11. Write a problem about feet and inches.

_____

_____

_____

Challenge

12. Each time George Grasshopper jumps, he moves twice as far as his last jump. His first jump is 3 inches. On which jump will he move 4 feet?   jump _____

Jump Number	1	2	3	4	5	6
Distance in Inches	3	6	12			

272 two hundred seventy-two   CCLXXII   136 + 136

Chapter 12 Lesson 5: Using Fractions to Measure Length

NCTM Standards 1, 2, 4, 6, 7, 8, 9, 10

How long is each piece of ribbon? Use a ruler to measure to the nearest half inch.

1.

   about $4\frac{1}{2}$ inches

2. about _____ inches

3. about _____ inches

4. about _____ inches

5. Draw a line that is about $3\frac{1}{2}$ inches long. Start at the dot.

   •

 NOTE: Your child is learning about half inches and fractions of a foot and of a yard.

91 + 91 + 91   CCLXXIII   two hundred seventy-three   273

Complete each table. What is missing?

6.
Number of Yards	$\frac{1}{3}$	$\frac{2}{3}$	1		3	4	
Number of Feet	1	2		6			15

7.
Number of Feet	$\frac{1}{12}$	$\frac{2}{12}$	$\frac{5}{12}$	$\frac{7}{12}$	$\frac{11}{12}$		2
Number of Inches	1	2				12	

Use the tables to complete the problems.

8. Tina cuts a 6-inch piece of ribbon. What fraction of a foot is the piece of ribbon?

    _____ of a foot

9. Bob jumps 24 inches. Kyle jumps 1 yard. Who jumped further?

    _____

10. Jill has 7 feet of ribbon. Amy has 2 yards of ribbon. Who has more ribbon? Use words, numbers, or pictures to explain.

Problem Solving

11. Can a piece of ribbon be 4 inches long when measured to the nearest inch and to the nearest half inch? Explain.

    _____

    _____

    _____

274   two hundred seventy-four   CCLXXIV   137 + 137

Chapter 12
Lesson 6: Measuring to the Nearest Centimeter

NCTM Standards 3, 4, 6, 9, 10

How long is the picture of each object? Use a ruler to measure to the nearest centimeter.

1.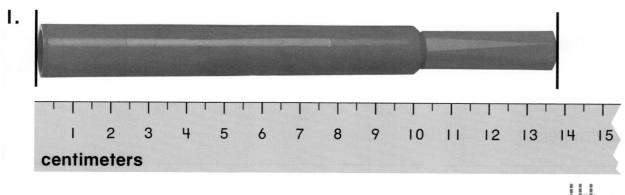

    about __14__ centimeters

2.

    about _____ centimeters

3.

    about _____ centimeters

4.

    about _____ centimeters

5.

    about _____ centimeter

6.

    about _____ centimeters

 NOTE: Your child is learning to measure length to the nearest centimeter.

CCLXXV two hundred seventy-five **275**

Draw a line for each length. Start at the dot.

7. • - - - -

    13 centimeters

8. •

    2 centimeters

9. •

    8 centimeters

Draw a picture of something in your classroom that matches each length.

10. about 5 centimeters
11. about 30 centimeters
12. about 45 centimeters

Challenge

Follow the clues to draw a figure.

13. The distance around a figure is 12 centimeters. The figure has 4 sides. The sides are NOT all the same length.

**Chapter 12
Lesson 7**

Measuring in Centimeters and Meters

NCTM Standards 1, 4, 6, 7, 8, 9, 10

Which unit would you use to measure each real object?
Circle *centimeters* or *meters*.

1.

centimeters
(meters)

2.

centimeters
meters

3.

centimeters
meters

4.

centimeters
meters

5.

centimeters
meters

6.

centimeters
meters

7.

centimeters
meters

8.

centimeters
meters

9.

centimeters
meters

NOTE: Your child is learning about centimeters and meters.

Choose a unit to measure each real object.
Then measure to the nearest unit.

Units of Measure
inches   feet   yards
centimeters   meters

Object	Estimate	Measurement
10. chalk		about _____
11. chalkboard		about _____
12. book		about _____

13. For Problem 12, how did you choose the unit? How did you measure the book? _____

_____

_____

Problem Solving

14. Kate and Carla each have a piece of string. Kate's string is 7 centimeters long. Carla's string is 7 inches long. Who has the longer string? _____

    Use words, numbers, or pictures to explain.

**Chapter 12
Lesson 8**

Problem Solving Strategy
Act It Out
NCTM Standards 1, 2, 4, 6, 7, 8, 9, 10

Understand
Plan
Solve
Check

1. A pencil case is 15 centimeters long.

   Can your pencil fit in this case? _____

   How long is your pencil? about _____

2. Pat's pencil is longer than this line.

   ▬▬▬▬▬▬▬▬▬▬▬▬

   How long might Pat's pencil be?   about _____

3. Tyler buys a book for 86¢. He pays the exact amount with the fewest coins. What coins does Tyler use?

   _____

4. Alice has 12 shells, Sandy has 9 shells and Wyatt has 15 shells. They want to share the shells equally. How many shells will each person get?

   _____ shells

NOTE: Your child is exploring different ways to solve problems. Sometimes acting it out is an efficient way to solve a problem.

93 + 93 + 93   **CCLXXIX**   two hundred seventy-nine **279**

Problem Solving Test Prep

1. When you double this odd number, you get a number between 50 and 55. What is the number?

   (A) 27
   (B) 29
   (C) 31
   (D) 53

2. There are 8 pages in a photo album. Each page has 6 photos on it. How many photos are in the whole album?

   (A) 14 photos
   (B) 24 photos
   (C) 36 photos
   (D) 48 photos

Show What You Know

3. Jan has already delivered 135 newspapers. He has 57 more newspapers to deliver. How many newspapers will Jan deliver in all?

   _____ newspapers

   Explain how you found the answer.

   _____
   _____
   _____
   _____

4. Two counters are tossed on this gameboard. Add the two scores. What total scores are possible?

   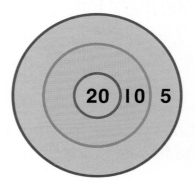

   _____

   Explain how you found the answer.

   _____
   _____
   _____
   _____
   _____

Chapter 12 Review/Assessment

NCTM Standards 1, 2, 3, 4, 6, 7, 8, 9, 10

1. Use a paper clip as the measuring unit.
   Draw a line 3 units long. Lesson 1

   •

2. How long is the picture of the crayon?
   Use a ruler to measure to the nearest inch. Lesson 2

   about _____ inches

3. Which unit would you use to measure a real swing set?
   Circle *inches, feet,* or *yards.* Lesson 3

   inches   feet   yards

4. Complete the table. Lesson 4

Number of Yards	1	2	3	4		6	7
Number of Feet		6			15	18	

100 + 100 + 81   CCLXXXI   two hundred eighty-one   281

5. How long is the piece of string? Use a ruler to measure to the nearest half inch. Lesson 5

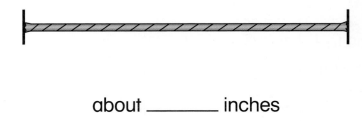

about _____ inches

6. How long is the picture of the ribbon? Use a ruler to measure to the nearest centimeter. Lesson 6

about _____ centimeters

7. Which unit would you use to measure the nail? Circle *centimeters* or *meters*. Lesson 7

centimeters   meters

Problem Solving Lesson 8

8. Reese has a piece of ribbon that is 2 feet long. She wants to cut it into 3 equal pieces. How many inches long is each piece?

_____ inches

Chapter 13

Name _____

Exploring Multiplication and Division
Choosing Snacks

You need
- milk containers
- juice boxes
- pieces of fruit
- pictures of fruits and drinks (optional)

Make and record different combinations.

STEP 1 Making Combinations

How many fruits do you have? _____

How many drinks do you have? _____

What one fruit did you choose? _____

What one drink did you choose? _____

STEP 2 Recording Combinations

What other combination of a drink and fruit could you make?

_____

Record all of the different combinations.

_____

_____

How many different combinations did you find? _____

STEP 3 Finding All Combinations

How do you know you found all the combinations?

_____

_____

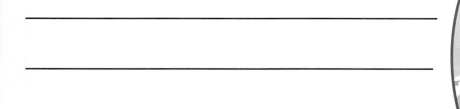

Investigation

School-Home Connection

Dear Family,

Today we started Chapter 13 of *Think Math!* In this chapter, I will explore combinations, intersections, and arrays as I learn about multiplication and division. There are NOTES on the Lesson Activity Book pages to explain what I am learning every day.

Here are some activities for us to do together at home. These activities will help me understand multiplication and division.

Love,

_____

Family Fun

What Will I Wear?

Work with your child to act out one of the activities from class.

- Use some of your child's clothing to find how many different outfits can be made. Take out 2 pairs of pants or skirts and 2 or 3 tops.

- Work with your child to make outfits by combining each top with a bottom. Help your child come up with a method to make sure you get all the combinations, such as pairing the first top with every bottom and then doing the same with each top.

- Together, make a list to keep track of all the different outfits. Count all of the different combinations.

- If time allows, add another top or bottom to see how many more outfits you can make.

Scavenger Hunt

Work with your child to count equal groups.

- Look around the house to find objects that are arranged in equal rows and columns. For example, tiles on the floor, pictures on a wall, panels on a door, or paints in a box.

- Help your child find the total number of objects in a display with equal groups. Talk about how to skip-count by the number of objects in a row or column. To find how many eggs are in a full carton, skip-count by twos or by sixes.

- Help your child write a multiplication sentence to show each arrangement.

$2 \times 6 = 12$ or $6 \times 2 = 12$

Chapter 13 Lesson 1: Counting Combinations

NCTM Standards 1, 2, 6, 7, 8, 9, 10

How many different outfits can be made each time?

1.
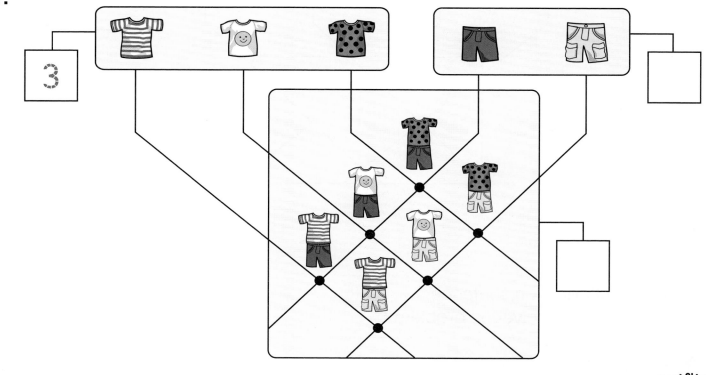

_____ outfits

2.
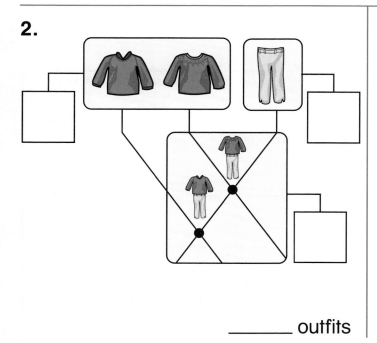

_____ outfits

3.
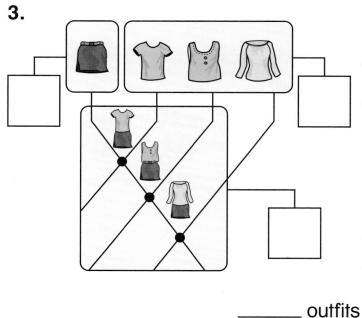

_____ outfits

NOTE: Your child is learning to find all possible combinations from two sets.

CCLXXXV two hundred eighty-five 285

4. Andrew has 2 shirts and 2 pairs of shorts for soccer. How many different uniforms can he make?

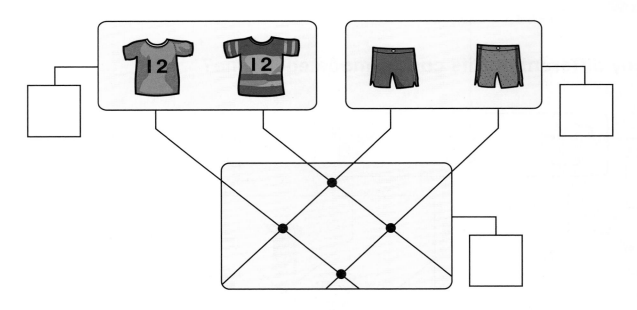

_____ uniforms

What multiplication sentence
can you use to solve the problem?   _____ × _____ = _____

 5. How many different outfits can you make from 3 shirts and 3 pairs of pants? Use words, numbers, or pictures to explain.

_____ outfits

Challenge

6. How many different two-digit numbers can you make using the digits 2, 3, and 4? List all of the numbers.

I can make _____ two-digit numbers.

Chapter 13 Lesson 2: Counting Intersections

NCTM Standards 1, 2, 3, 6, 7, 8, 9, 10

How many intersections are there? Write the missing numbers.

1.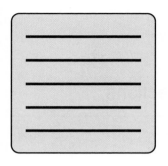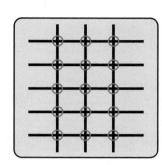

   _5_ lines   _____ lines   _____ intersections

   $5 \times 3 =$ _____

   $3 \times 5 =$ _____

2.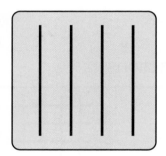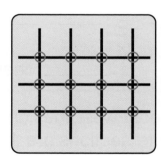

   _____ lines   _____ lines   _____ intersections

   $3 \times 4 =$ _____

   $4 \times 3 =$ _____

3. Draw the intersections.

    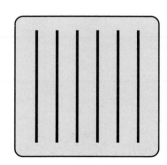 

   _____ lines   _____ lines   _____ intersections

   $2 \times 6 =$ _____

   $6 \times 2 =$ _____

NOTE: Your child is working with intersecting lines to learn multiplication facts.

200 + 80 + 7   CCLXXXVII   two hundred eighty-seven   **287**

What is missing? Draw lines and numbers to show the multiplication.

4.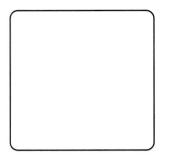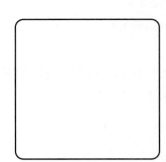

$2 \times 4 = $ _____

$4 \times 2 = $ _____

_____ lines   _____ lines   _____ intersections

5.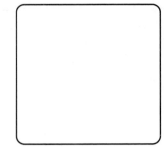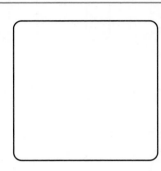

$3 \times 3 = $ _____

_____ lines   _____ lines   _____ intersections

 6. There are 4 east-west lines and 4 north-south lines. How many intersections are there? Explain how you found the answer.

_____

_____

_____

Problem Solving

7. In Bridgetown, 4 streets go north-south. Every north-south street crosses every east-west street. There is a stoplight at every intersection. There are 24 stoplights in town. How many streets go east-west?

_____ east-west streets

288 two hundred eighty-eight   CCLXXXVIII   24 dozen

Chapter 13
Lesson 3

Counting Hidden Intersections

NCTM Standards 1, 2, 3, 6, 8, 9, 10

How many intersections are there?
Write the missing numbers.

1.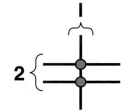

   $2 \times 1 = \underline{2}$

   $1 \times 2 = \underline{2}$

2.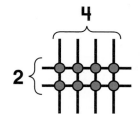

   $2 \times 4 = \underline{}$

   $4 \times 2 = \underline{}$

3.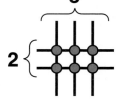

   $3 \times 2 = \underline{}$

   $2 \times 3 = \underline{}$

4.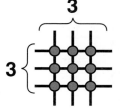

   $3 \times 3 = \underline{}$

5.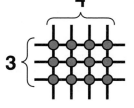

   $4 \times 3 = \underline{}$

   $\underline{} \times \underline{} = \underline{}$

6.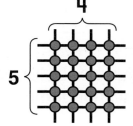

   $4 \times 5 = \underline{}$

   $\underline{} \times \underline{} = \underline{}$

7.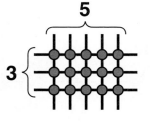

   $\underline{} \times \underline{} = \underline{}$

   $\underline{} \times \underline{} = \underline{}$

8.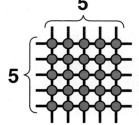

   $\underline{} \times \underline{} = \underline{}$

NOTE: Your child is working with intersecting lines to learn multiplication facts.

290 – 1   CCLXXXIX   two hundred eighty-nine

How many intersections are there?
Write the missing numbers.

9.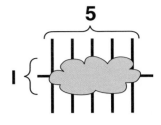

___ × ___ = ___

___ × ___ = ___

10.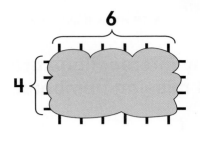

___ × ___ = ___

___ × ___ = ___

11.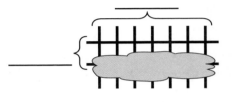

___ × ___ = ___

___ × ___ = ___

12.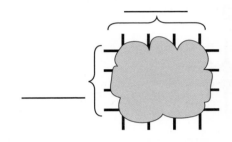

___ × ___ = ___

13. Choose an even number between 10 and 30. Draw a town map with that many intersections.

___ intersections

Problem Solving

14. Bear Town has 8 intersections. Every north-south street crosses every east-west street. How many streets could there be?
Explain how you found the answer.

Draw streets to make 8 intersections. Count the streets.

___ streets

_____

_____

Chapter 13
Lesson 4 — Introducing Division
NCTM Standards 1, 2, 3, 6, 8, 9, 10

Write the missing numbers.

1.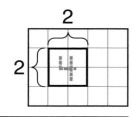

2.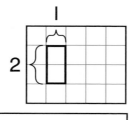

3.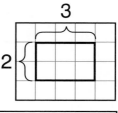

4.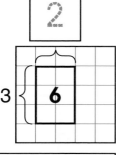

5.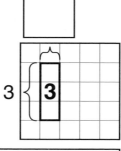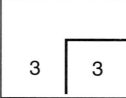

6.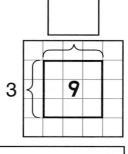

7.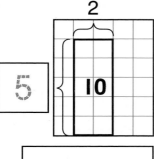

8.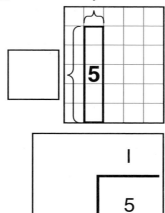

9.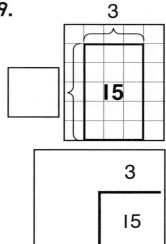

NOTE: Your child is learning to divide by thinking about missing factors.

97 + 97 + 97   CCXCI   two hundred ninety-one   291

Write the missing numbers.

10. 11. 12.

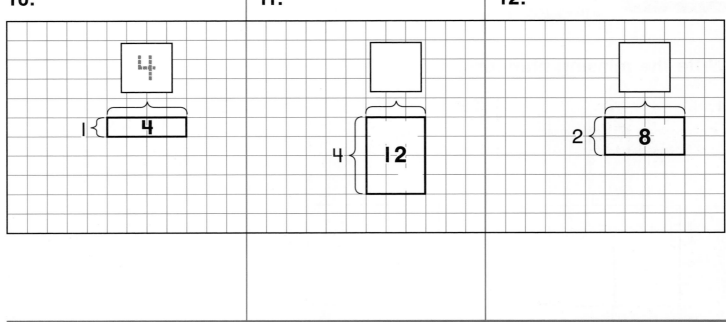

Here is a shorter way to write the examples for Problems 10 to 12. Write the missing numbers.

13. 14. 15.

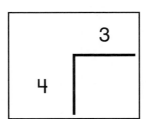

Challenge

16. What is the missing number? Tell how you know.

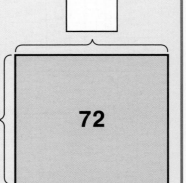

Chapter 13 Lesson 5
Multiplication and Division Fact Families

NCTM Standards 1, 2, 3, 6, 8, 9, 10

Complete each fact family.

1.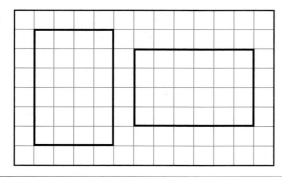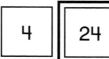

   4 × 6 = 24   6 × 4 = ___

2.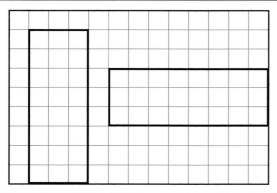

   8 × ___ = 24   3 × ___ = 24

3.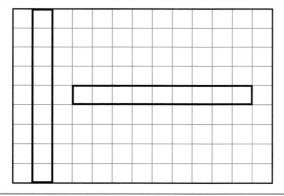

   9 × 1 = ___   1 × 9 = ___

4.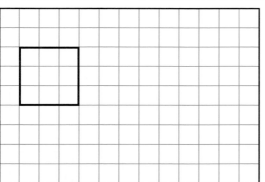

   ___ × ___ = ___   ___ × ___ = ___

NOTE: Your child is learning to write all members of multiplication and division fact families.

300 − 7   CCXCIII   two hundred ninety-three   293

5. Some friends share 8 cookies. Write the missing numbers. Show the fact family for the pictures.

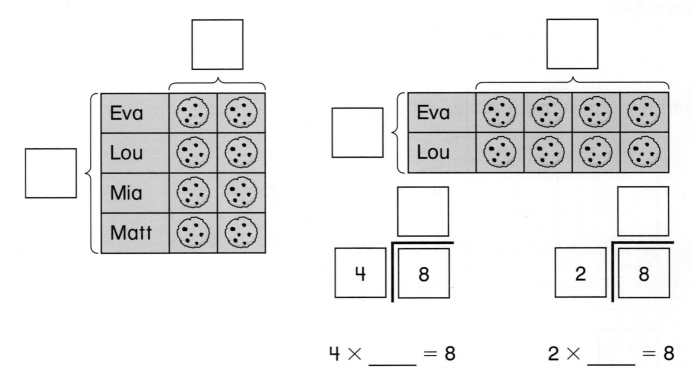

4 × ____ = 8   2 × ____ = 8

6. Draw a rectangle on the grid. Write the fact family for the rectangle.

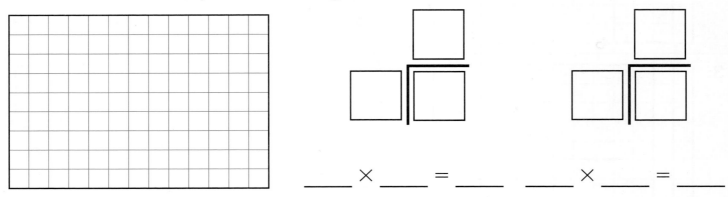

____ × ____ = ____   ____ × ____ = ____

Challenge

7. Write a fact family for a rectangle with 20 squares

____ × ____ = ____

____ × ____ = ____

Chapter 13 Lesson 6: Multiplication and Division Models

NCTM Standards 1, 2, 6, 7, 8, 9, 10

**Complete each model.
What are the missing numbers?**

1. $3 \times 5 =$ ___

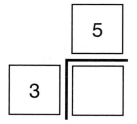 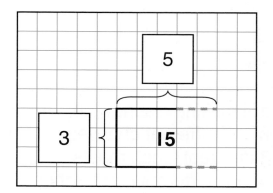

2. $7 \times 3 =$ ___

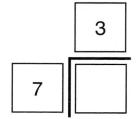 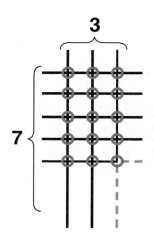

3. $2 \times 6 =$ ___

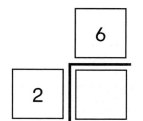 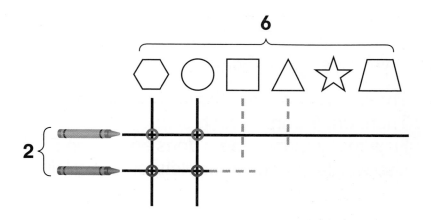

NOTE: Your child is learning basic multiplication facts by working with different models.

CCXCV two hundred ninety-five 295

How can you solve each problem? Show your work.

4. Two children equally share a pack of 18 stickers. How many stickers does each child get?

   _____ stickers

5. Callie is making a sandwich. She has 2 different cheeses and 3 different lunch meats. How many different sandwiches of one meat and one cheese can she make?

   _____ sandwiches

6. Five children go to the fair. Each child wins 5 goldfish. How many goldfish do they win in all?

   _____ goldfish

Problem Solving

7. Three friends share 2 boxes of granola bars. There are 9 bars in each box. How many granola bars does each friend get? Use words, numbers, or pictures to explain.

   _____ bars

Chapter 13
Lesson 7

Dividing and Estimating with Coins

NCTM Standards 1, 2, 6, 7, 8, 9, 10

**How many coins make one dollar?
Write the missing numbers.**

1.

   1¢

   _100_ pennies = $1.00

   ____ × 1 = 100

2.

   5¢

   ____ nickels = $1.00

   ____ × 5 = 100

3.

   10¢

   ____ dimes = $1.00

   ____ × 10 = 100

4.

   25¢

   ____ quarters = $1.00

   ____ × 25 = 100

5.

   50¢

   ____ half dollars = $1.00

   ____ × 50 = 100

6.

   100¢

   ____ dollar coin = $1.00

   ____ × 100 = 100

NOTE: Your child is learning to work with groups of coins equal to whole dollar amounts. Together, find how many dimes are equal to $2.00.

99 + 99 + 99   CCXCVII   two hundred ninety-seven   **297**

Circle the best estimate for each problem.

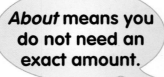
About means you do not need an exact amount.

7. You have 32 dimes. About how many dollars do you have?

      $1.00 $2.00   ($3.00)

8. You have 25 quarters. About how many dollars is that?

    $2.00 $6.00 $8.00

9. You have 21 half dollars. About how many dollars is that?

    $5.00 $9.00 $10.00

 10. You have $4.00 in one kind of coin. How many coins might you have? Explain.

_____

_____

_____

_____

Problem Solving

11. Jose is saving quarters. He gets one quarter each day. How many days will he need to save for a total of $2.00?

    _____ days

Chapter 13
Lesson 8

Problem Solving Strategy
Guess and Check
NCTM Standards 1, 2, 3, 6, 7, 8, 9, 10

Understand
Plan
Solve
Check

1. My town has 21 intersections. There are 10 roads. How many north-south roads and east-west roads can my town have? Explain how you found the answer.

   _____ north-south roads

   _____ east-west roads

   _____
   _____
   _____

2. Three children equally share 27 pennies. How many pennies does each child get? Explain.

   _____ pennies

   _____
   _____
   _____

3. Jodie has 2 different pair of pants. She can make 14 outfits. How many shirts does Jodie have? Explain.

   _____ shirts

   _____
   _____
   _____

 NOTE: Your child is exploring different ways to solve problems. Sometimes using the strategy, *guess and check*, is an efficient way to solve a problem.

300 − 1   **CCXCIX**   two hundred ninety-nine **299**

Problem Solving Test Prep

1. Jeff has a penny, a nickel, a dime, and a quarter. He picks two coins. Which is NOT an amount of money he could have?

   Ⓐ 15¢

   Ⓑ 21¢

   Ⓒ 26¢

   Ⓓ 35¢

2. A snail travels 2 inches every hour. If it starts moving at 3 o'clock, how far would it get by 5 o'clock?

   Ⓐ 2 inches

   Ⓑ 3 inches

   Ⓒ 4 inches

   Ⓓ 5 inches

Show What You Know

3. Bob has a pack of 16 batteries. He puts the same number of batteries in each toy car. He has enough batteries to get 5 cars running. How many batteries go in each car?

   _____ batteries

   Explain your answer.

   _____

   _____

   _____

   _____

   _____

4. Dana has a rectangular piece of cloth. She cuts the cloth with 4 straight lines to get all triangle pieces. How many triangles does she make?

   _____ triangles

   Explain your answer.

   _____

   _____

   _____

   _____

   _____

three hundred   CCC   60   25 dozen

Chapter 13 Review/Assessment

NCTM Standards 1, 2, 3, 4, 6, 7, 8, 9, 10

1. How many different outfits can you make? Lesson 1

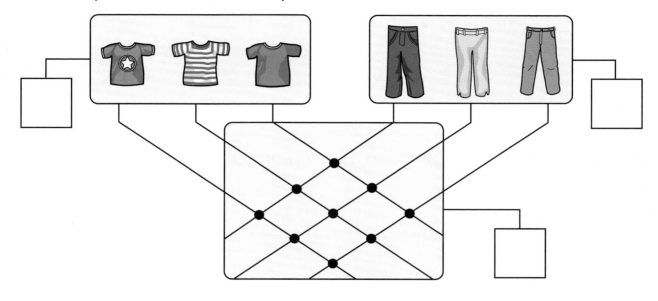

_____ outfits

2. What is missing? Draw lines and numbers to show the multiplication. Lesson 2

 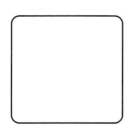 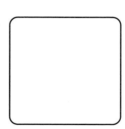

$3 \times 4 =$ _____

$4 \times 3 =$ _____

_____ lines   _____ lines   _____ intersections

Write the missing numbers. Lesson 4

3.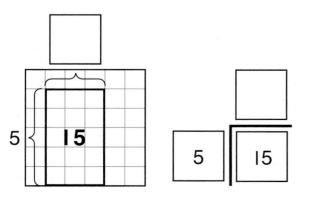

4.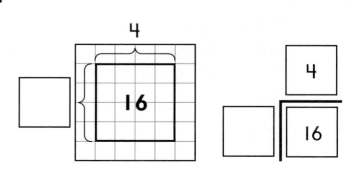

300 + 1   **CCCI**   three hundred one   301

5. Complete the fact family. Lesson 5

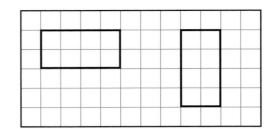 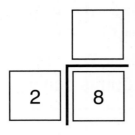 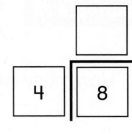

$2 \times \underline{} = 8 \qquad 4 \times \underline{} = 8$

6. Make a model to solve the problem. Show your work. Lesson 6

Six friends go to a book fair.
Each friend buys 4 books.
How many books do they buy in all?

_____ books

7. You have 28 dimes. About how many dollars do you have? Circle the best estimate. Lesson 7

      $2.00 $3.00 $28.00

Problem Solving Lesson 8

8. Tyson has 20 baseball cards in a book. One page holds 6 cards. How many pages of the book are full?

_____ pages

Chapter 14

Name _____

Comparing and Contrasting Three-Dimensional Figures
Slide, Stack, and Roll

You need
- Three-dimensional blocks

What are some ways you can describe a figure?

STEP 1 Exploring a Block

Describe your block.

_____

_____

STEP 2 Sliding, Stacking, and Rolling

Draw each block and put a ✓ if it can slide, stack, or roll.

Figure	Slide	Stack	Roll
□			
○			
⬭			

STEP 3 Comparing Blocks

Compare blocks. Describe the blocks that could roll. How are they different from the other blocks?

_____

_____

School-Home Connection

Dear Family,

Today we started Chapter 14 of *Think Math!* In this chapter, I will explore three-dimensional figures and learn how they are alike and different. There are NOTES on the Lesson Activity Book pages to explain what I am learning every day.

Here are some activities for us to do together at home. These activities will help me as I learn about three-dimensional figures.

Love,

_____

Family Fun

What's My Figure?

Work with your child to play this game. Your child will play this game in Lesson 1.

- Think of an object in the house that is shaped like one of the following figures.

cube   rectangular prism   sphere   prism

For example, a tissue box shaped like a rectangular prism.

- Have your child ask yes/no questions to try and guess your secret object. Some possible questions might be:

  Is there a triangle in the figure?

  Does it roll?

- Have your child continue asking questions until he or she has correctly identified the object.

- Switch roles and play again.

Making Faces

Together, trace a figure to see the faces.

- You will need a three-dimensional object and a pencil and paper.

- Help your child trace around one of the sides of the object.

- Ask your child to name the figure you traced.

- Try other objects. Talk about what sides you can trace and what shapes you will make.

Chapter 14
Lesson 1

Two- and Three-Dimensional Figures

NCTM Standards 2, 3, 6, 8, 9, 10

Match each object to a set of three-dimensional figures.

1.

2.

3.

4.

5.

6.

7.

8.

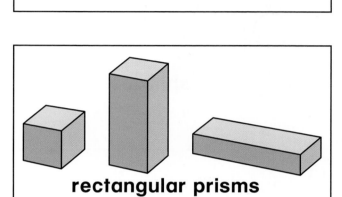
spheres

rectangular prisms

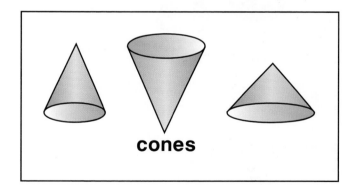
cones

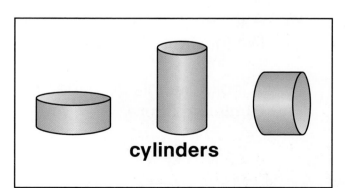
cylinders

NOTE: Your child is learning about three-dimensional figures. Have your child look for objects around your home to match the three-dimensional figures on this page.

CCCV three hundred five 305

Which figure does not belong?

9.

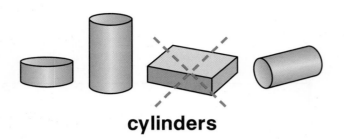

cylinders

10.

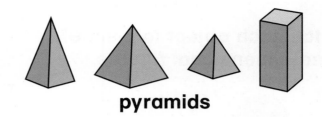

pyramids

11.

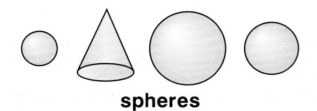

spheres

12.

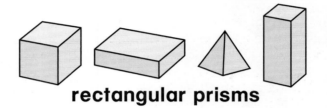

rectangular prisms

13.

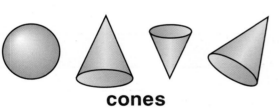

cones

14. Draw an object that does not belong in this set.

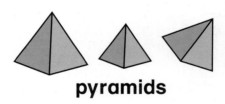

pyramids

Challenge

15. A three-dimensional figure looks like this from the side.

It looks like this from the bottom. Circle the figure.

Chapter 14 Lesson 2 — Faces

NCTM Standards 2, 3, 6, 7, 8, 9, 10

Match each three-dimensional figure to its faces.

1.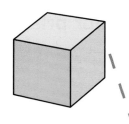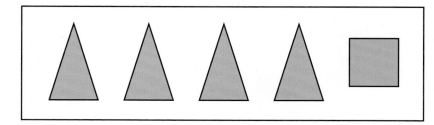

2.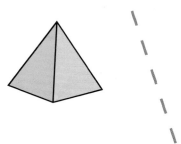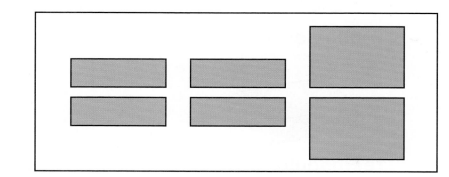

3.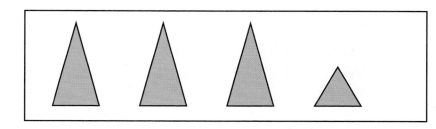

4.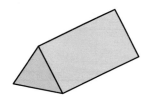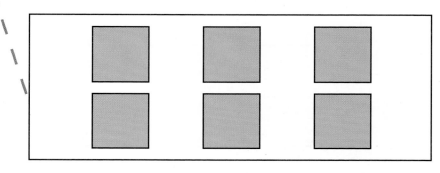

5.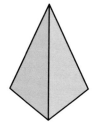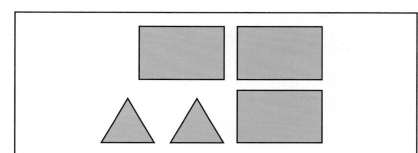

 NOTE: Your child is learning about the faces of three-dimensional figures.

290 + 17   **CCCVII**   three hundred seven   **307**

6. Circle each figure with 6 faces. Put an X on each figure with 5 faces. Put a ☐ on each figure with 0 faces.

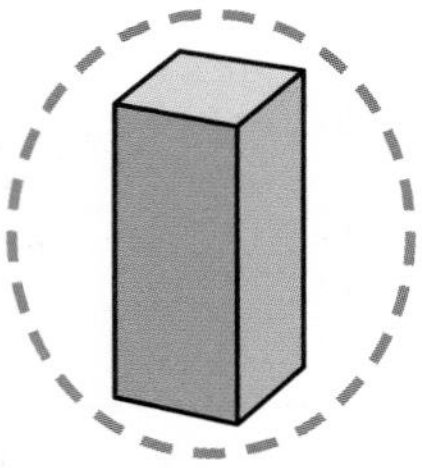

rectangular prism 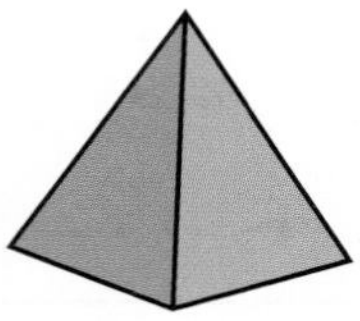   square pyramid    rectangular prism

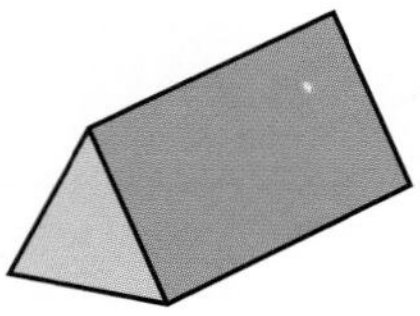

triangular prism    cube 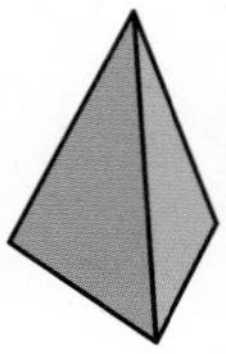   triangular pyramid

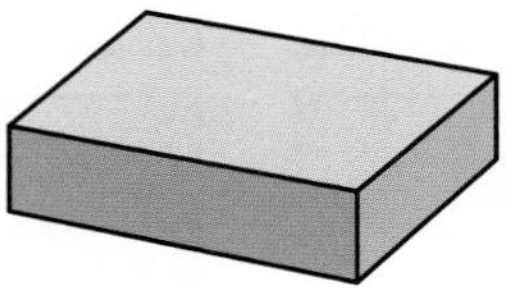

rectangular prism 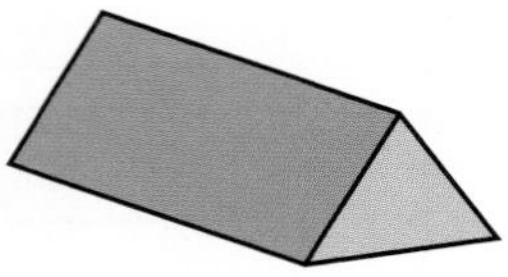   triangular prism 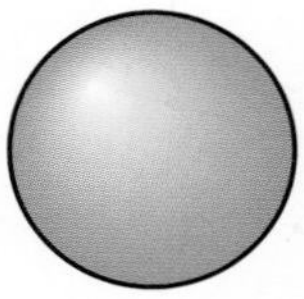   sphere

---

7. Look at a triangular prism and a triangular pyramid. What is the same about them? What is different?

_____

_____

_____

_____

Challenge

8. What three-dimensional figure has faces in both of these shapes?

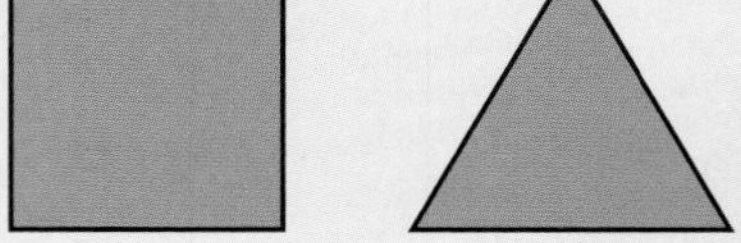

_____

Chapter 14 Lesson 3 **Edges**

Name _____ Date/Time _____

NCTM Standards 2, 3, 6, 7, 8, 9, 10

I built these models with toothpicks and marshmallows.

Match each three-dimensional figure to its model. How many toothpicks do you need to make each model?

1.
cube

2.
triangular pyramid

3.
rectangular prism

4.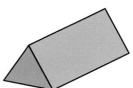
triangular prism

5.
square pyramid

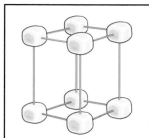
_____ toothpicks

_____ toothpicks

_____ toothpicks

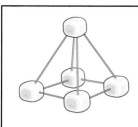
_____ toothpicks

_____ toothpicks

 NOTE: Your child is learning about the edges of three-dimensional figures.

103 + 103 + 103   CCCIX   three hundred nine   **309**

How many edges does each object have?

	Object	Number of Edges
6.	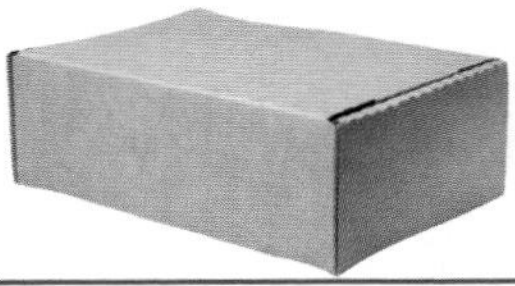 (box)	
7.	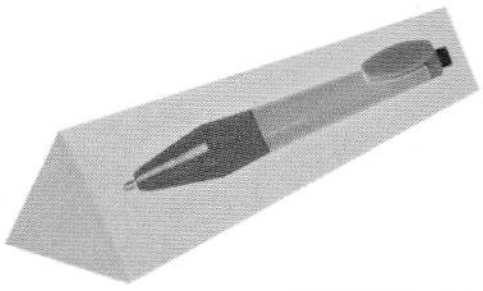 (triangular prism)	
8.	 (tissue box)	
9.	 (globe)	

10. Draw your own object.

11. Explain your answer to Problem 9.

_____

_____

_____

Problem Solving

12. The edge of a cube is 2 inches long. How much yarn would you need to cover every edge of the cube? Use words, numbers, or pictures to explain.

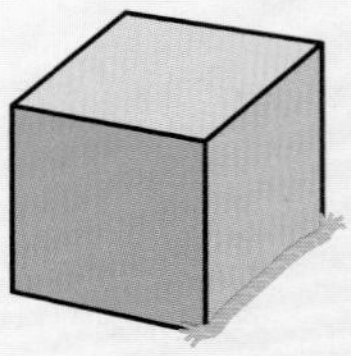

    _____ of yarn

Chapter 14
Lesson 4: Vertices

NCTM Standards 2, 3, 6, 7, 8, 9, 10

Match each three-dimensional figure to its description.

1.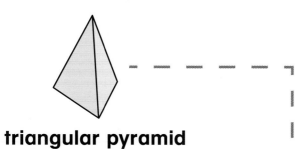
triangular pyramid

   6 faces, 12 edges, 8 vertices

2.
rectangular prism

   5 faces, 9 edges, 6 vertices

3.
square pyramid

4.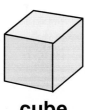
cube

   4 faces, 6 edges, 4 vertices

5.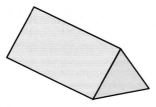
triangular prism

   5 faces, 8 edges, 5 vertices

NOTE: Your child is learning about the vertices of three-dimensional figures.

300 + 10 + 1   **CCCXI**   three hundred eleven   **311**

How many faces, edges, and vertices does each object have?

	Object	Number of Faces	Number of Edges	Number of Vertices
6.				
7.				
8.				
9.				
10.	Find your own object.			

Challenge

How many faces, edges, and vertices are in the blocks that make up each building?

11.   _____ faces   _____ edges   _____ vertices

12. 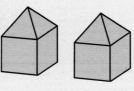   _____ faces   _____ edges   _____ vertices

312 three hundred twelve CCCXII 104 + 104 + 104

Chapter 14 Lesson 5: Cylinders and Cones

NCTM Standards 2, 3, 6, 7, 8, 9, 10

Which figure in each row could have made the footprint?

Footprint	Figures
1. circle	cube, pyramid, triangular prism, **cone** (circled)
2. triangle	cylinder, cone, pyramid, cube
3. square	pyramid, cylinder, cone, cylinder (on side)
4. circle	cube, rectangular prism, pyramid, cylinder
5. rectangle	sphere, triangular prism, pyramid, cone
6. circle	cone, cube, rectangular prism, pyramid

NOTE: Your child is learning about cylinders and cones. Together find items in the house that look like a cylinder or a cone.

Match each figure to its attribute.

7.
triangular pyramid

a. 1 vertex

8.
rectangular prism

b. 6 faces

9.
triangular prism

c. 6 edges

10.
cone

d. 5 faces

11.
cylinder

e. 2 faces

Problem Solving

12. What three-dimensional figure can you make from this picture? Tell how you know.

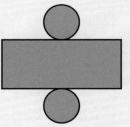

_____

_____

_____

Chapter 14 Lesson 6

Problem Solving Strategy
Make a Table

NCTM Standards 1, 2, 3, 6, 7, 8, 9, 10

Understand
Plan
Solve
Check

1. Bob has 5 triangular prisms. He paints the rectangle faces blue and the triangle faces red. How many blue and red faces does he paint?

Number of Prisms	1	2	3	4	5
Rectangle Faces	3				
Triangle Faces	2				10

_____ blue faces   _____ red faces

2. Jana saves $2 this week. Each week she saves double the amount from the week before. How much money will she save after 4 weeks?

_____

Week	Amount Saved
1	$2
2	$4

3. An octopus has 8 arms. How many arms are on 7 octopuses?   _____ arms

Number of Octopuses	1						
Number of Arms	8						

NOTE: Your child is exploring different ways to solve problems. Sometimes making a table is an efficient way to solve a problem.

CCCXV   three hundred fifteen   **315**

Problem Solving Test Prep

1. Hope reads every night for 20 minutes. Tonight she finishes at 8:10. What time did she start reading?

    Ⓐ 7:40

    Ⓑ 7:50

    Ⓒ 8:20

    Ⓓ 8:30

2. Piedad starts a number pattern with 3, 7, 11, and 15. If she continues this pattern, what will be the eighth number?

    Ⓐ 17

    Ⓑ 19

    Ⓒ 27

    Ⓓ 31

Show What You Know

3. Jim multiplies a number by itself. The number he gets is between 20 and 30. What is the number?

    _____

    Explain how you found the answer.

    _____

    _____

    _____

    _____

4. Craig has 12 stickers, Jean has 17 stickers, and Cliff has 13 stickers. They want to share the stickers equally. How many stickers will each person get?

    _____ stickers

    Explain how you found the answer.

    _____

    _____

    _____

    _____

    _____

Chapter 14 Review/Assessment

NCTM Standards 1, 2, 3, 4, 6, 7, 8, 9, 10

Put an X on the figure that does not belong. Lesson 1

1.
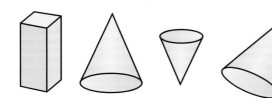

2.
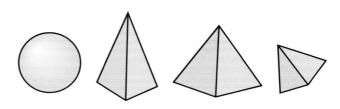

3. Circle each figure with 5 faces.
   Put an X on each figure with 6 faces. Lesson 2

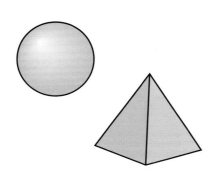
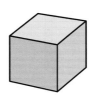
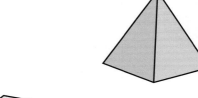
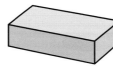

4. Look at a cube and a square pyramid.
   What is the same about them?
   What is different? Lessons 2, 3

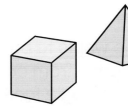

_____

_____

_____

_____

_____

300 + 10 + 7   CCCXVII   three hundred seventeen **317**

How many faces, edges, and vertices does each figure have? Lessons 2–4

	Figure	Number of Faces	Number of Edges	Number of Vertices
5.				
6.				
7.				

Which figure in each row could have made the footprint? Lesson 5

	Footprint	Figures
8.		
9.		

Problem Solving

10. Eric wants to build a square pyramid and a triangular pyramid from cut-out shapes. How many square faces and triangle faces does he need? Lesson 6

Figure	Square Faces	Triangle Faces
Square Pyramid		
Triangular Pyramid		
Total		

Chapter 15

Name _____

Capacity, Weight/Mass, and Temperature
Measuring Different Attributes

You need
- coffee mug
- ruler, pan balance, paper cup, water, rice, thermometer

What are all of the different ways that you can measure a mug?

STEP 1 Thinking About Attributes

What can you measure about a mug?

_____
_____

STEP 2 Working with Tools

How might you use tools to measure a mug? _____
_____
_____
_____

STEP 3 Measuring in Different Ways

Use the tools to measure a mug in different ways. What did you find out?

_____
_____
_____

Investigation

School-Home Connection

Dear Family,
Today we started Chapter 15 of *Think Math!* In this chapter, I will explore how to measure capacity, weight/mass, and temperature. There are NOTES on the Lesson Activity Book pages to explain what I am learning every day.

Here are some activities for us to do together at home. These activities will help me understand measurement.

Love,

_____

Family Fun

What Is the Temperature?

Work with your child to record the daily air temperature.

- Each day for a week, look in the newspaper to find and record the daily forecast of high and low temperatures.

- Look at your list of temperatures for the week. Talk about which day was the warmest and which was the coldest. Talk about what clothes you might wear in these different temperatures.

- With your child, make a prediction for tomorrow's high and low temperatures.

- Continue recording temperatures beyond the week, if you wish. Then use this information to help your child plan his or her clothes or activities.

Market Measures

Work with your child to identify units of measure at the store.

- Together, look at flyers for the supermarket. Then take a visit to the store and look at different products on the shelves.

- Talk about the weights and liquid measures of meats, juices, cereal, and other products.

- Compare prices of products in different sizes. Decide which is the best buy.

- Have fun shopping and learning together!

Chapter 15 Lesson 1

Comparing, Ordering, and Measuring Capacity

NCTM Standards 2, 4, 6, 7, 8, 9, 10

Which unit is best to measure the capacity of each container?

spoon　**cup**　**pail**

Choose the unit that makes the most sense.

1.

   spoons　(cups)　pails

2.

   spoons　cups　pails

3.

   spoons　cups　pails

4.

   spoons　cups　pails

5.

   spoons　cups　pails

6.

   spoons　cups　pails

7.

   spoons　cups　pails

8.

   spoons　cups　pails

 NOTE: Your child is learning to compare, order, and measure containers by how much they hold.

107 + 107 + 107　　**CCCXXI**　　three hundred twenty-one　**321**

9. Becky has a bowl of soup.
Kari has a cup of soup.
Who has more soup?

_____

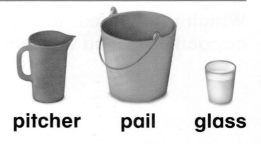

bowl   cup

10. Kate has a pitcher of water. Bud has a pail of water. Max has a glass of water. Who has the most water?

_____

pitcher   pail   glass

11. The metal vase holds 10 cups of water.
The glass vase holds 10 spoonfuls.
Which vase holds less water?

_____

metal vase   glass vase

12. How can you find out which of two bowls holds more?

_____

_____

Problem Solving

13. Ray has a red, a blue, and a green pail. The blue pail holds the most. The green pail holds more than the red. Which color pail holds the least? Tell how you found the answer.

_____

_____

_____

322 three hundred twenty-two   CCCXXII   161 + 161

Chapter 15 Lesson 2
Measuring in Cups, Pints, Quarts, and Gallons

NCTM Standards 1, 2, 4, 6, 8, 9, 10

Each baby lamb drinks 1 cup of milk. Match the lambs to what they drink.

2 cups = 1 pint
2 pints = 1 quart
4 quarts = 1 gallon

1.

1 pint

2.

3 pints

3.

1 cup

4.

2 quarts

5.

3 cups

6.

1 quart

NOTE: Your child is learning about cups, pints, quarts, and gallons. Show your child an empty milk container and ask him or her to find out how many cups will fill the container.

Which measurement from the box solves each riddle?

7. I am more than 1 cup. I am less than 3 cups. What am I?

_____ 1 pint _____

Measurements
1 pint   3 quarts
1 quart 2 gallons
1 gallon

8. I am more than 3 cups. I am less than 5 cups. What am I?

_____

9. I am more than 4 pints. I am less than 1 gallon. What am I?

_____

10. I have the same capacity as 16 cups. What am I?

_____

11. I have the same capacity as 8 quarts. What am I?

_____

 12. Make up your own riddle.

_____

_____

_____

Problem Solving

13. Tina has a 1-gallon punch bowl. She wants to fill the bowl with orange juice and lemonade. How much of each can she use?

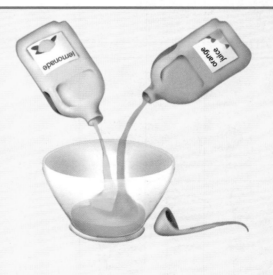

_____ orange juice   _____ lemonade

324   three hundred twenty-four   CCCXXIV   27 dozen

Name _____ Date/Time _____

Chapter 15 Lesson 3 — Measuring in Milliliters and Liters

NCTM Standards 1, 4, 6, 7, 8, 9, 10

Which unit is better to measure the capacity of each object?

1.

   (milliliters)   liters

2.

   milliliters   liters

3.

   milliliters   liters

4.

   milliliters   liters

5.

   milliliters   liters

6.

   milliliters   liters

Draw something you might measure with each unit.

7.

milliliters

8.

liters

NOTE: Your child is learning about milliliters and liters. Ask your child to guess and then check to find out if various household containers hold more than, less than, or the same as a liter container.

65   CCCXXV   three hundred twenty-five   325

9. Anne drank a glass of milk. Did she drink 230 milliliters or 230 liters of milk?

_____

10. Jeff filled a watering can with water. Did he use 3 liters or 3 milliliters of water?

_____

11. Todd ate a spoonful of soup. Did he eat 15 milliliters or 15 liters of soup?

_____

12. Nina filled the kitchen sink with water. Did she use 25 milliliters or 25 liters of water?

_____

✏️ 13. Write your own problem. Have a classmate solve it.

_____

_____

_____

Problem Solving

14. The gas tank in Mr. Brown's car holds about 60 liters. Does the gas tank hold more than or less than 60 gallons? Explain how you know. _____

_____

_____

Chapter 15 Lesson 4 — Comparing and Measuring Weight

NCTM Standards 2, 4, 6, 7, 8, 9, 10

Which unit is best to measure the weight of each object?

paper clip   book   brick

1.

(paper clips)   books   bricks

2.

paper clips   books   bricks

3.

paper clips   books   bricks

4.

paper clips   books   bricks

5.

paper clips   books   bricks

6.

paper clips   books   bricks

 7. Why did you choose that unit for Problem 6? Use words, numbers, or pictures to explain.

NOTE: Your child is learning to compare and measure objects by how much they weigh.

109 + 109 + 109   CCCXXVII   three hundred twenty-seven   **327**

 8. Ben's apple weighs more than Ali's apple. Ali's apple weighs more than Casey's apple. Can you tell whose apple weighs the most? Explain.

_____
_____
_____

 9. Ali's book weighs more than Casey's book. Ben's book weighs more than Casey's book. Can you tell whose book weighs the most? Explain.

_____
_____
_____

Challenge

10. Label the bags in order from lightest to heaviest. Write A, B, and C.

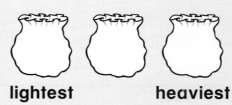

lightest   heaviest

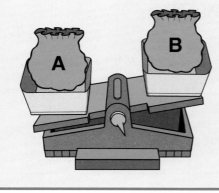
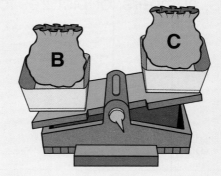

Chapter 15 Lesson 5: Measuring in Grams and Kilograms

NCTM Standards 1, 2, 4, 6, 8, 9, 10

What is missing? Complete the table.

	Object	More than or less than 1 gram?	More than or less than 1 kilogram?
1.	marker	more than	
2.	postage stamp		
3.	computer		
4.		more than	less than
5.		more than	more than
6.		less than	

NOTE: Your child is learning to estimate and measure objects in grams and kilograms. Ask your child to name some kitchen objects that are more than 1 kilogram.

300 + 20 + 9   CCCXXIX   three hundred twenty-nine   329

Draw two objects from your classroom. Estimate each one in grams or kilograms. Then measure.

	Object	Estimate	Measurement
7.		about _____	about _____
8.		about _____	about _____

9. Find two objects in your classroom that are each about 1 kilogram. Draw them.

Problem Solving

10. An adult cocker spaniel is about 12 kilograms. Would a cocker spaniel puppy measure 3 kilograms or 30 kilograms? Tell how you know.

_____

_____

_____

330 three hundred thirty

Chapter 15 Lesson 6

Measuring in Ounces, Pounds, and Tons

NCTM Standards 1, 4, 6, 9, 10

Which unit is best to weigh each object?

1.
ounces (pounds) tons

2.
ounces pounds tons

3.
ounces pounds tons

4.
ounces pounds tons

5.
ounces pounds tons

6.
ounces pounds tons

Draw something you might weigh with each unit.

7.

ounces

8.

pounds

9.

tons

 NOTE: Your child is learning about ounces, pounds, and tons. Ask your child to estimate and weigh grocery items.

150 + 150 + 31   CCCXXXI   three hundred thirty-one **331**

Match each animal to its weight.

10.  whale   12 pounds

11.  lion   10 ounces

12. cat   125 tons

13. kitten   400 pounds

14. Draw a classroom object that you think weighs between 1 pound and 3 pounds. Estimate and measure the weight.

   Estimate: about _____ pounds

   Measurement: about _____ pounds

Challenge

15. Together, 5 identical marbles weigh 1 ounce. How many marbles would you need to balance a 5-ounce fork?

   _____ marbles

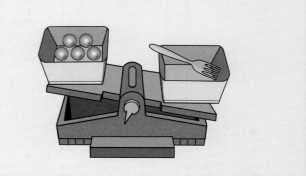

Chapter 15 Lesson 7

Name _____ Date/Time _____

Measuring Temperature

NCTM Standards 1, 2, 4, 6, 8, 9, 10

What temperature goes with each picture?
Color the thermometer to show your estimate.

1.

_____ °F

2.

_____ °F

3.

_____ °F

4. Draw your own.

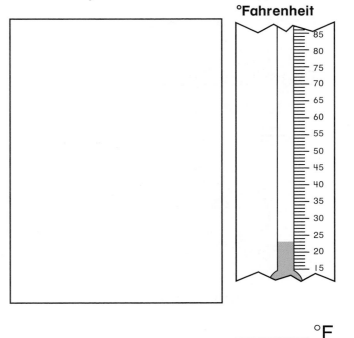

_____ °F

 NOTE: Your child is learning how to estimate and measure temperature in degrees Fahrenheit. Together, read and record the outside temperature for a week or so using your home thermometer.

III + III + III   CCCXXXIII   three hundred thirty-three   **333**

Play the cold version of *What's My Temperature?*
Pick a secret temperature between 25°F and −10°F.
Your partner asks *yes/no* questions. Use red and blue markers to record on the gameboard below.

Cold!

What's My Temperature? Gameboard

Don't forget! Use red if the guess is too hot. Use blue if it is too cold.

**Chapter 15
Lesson 8**

Problem Solving Strategy
Act It Out

NCTM Standards 1, 2, 4, 5, 6, 7, 8, 9, 10

Understand
Plan
Solve
Check

1. Shayna put 15 cubes in two bags. She put more cubes in Bag B than in Bag A. She balanced the pans by adding 3 cubes to the side with Bag A. How many cubes were in each bag?

   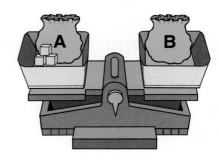

   Bag A _____ cubes   **Bag B** _____ cubes

2. Pat put 13 cubes in two bags. He put more cubes in Bag A than in Bag B. He balanced the pans by adding 5 cubes to the side with Bag B. How many cubes were in each bag?

   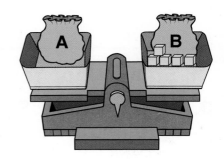

   Bag A _____ cubes   **Bag B** _____ cubes

3. Deion put 10 cubes in three bags. He put the same number of cubes in Bags A and B. He put more cubes in Bag C than in the other two bags together. He balanced Bags B and C by adding 4 cubes to the side with Bag B. How many cubes were in each bag?

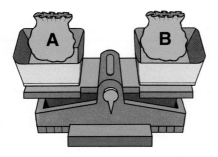 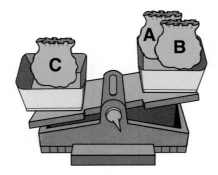 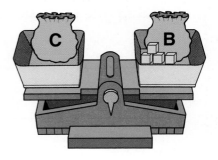

Bag A _____ cubes   **Bag B** _____ cubes   **Bag C** _____ cubes

NOTE: Your child is exploring different ways to solve problems. Sometimes acting it out is an efficient way to solve a problem.

67   CCCXXXV   three hundred thirty-five   **335**

Problem Solving Test Prep

1. I have 6 faces, 12 edges, and 8 vertices. All of my faces are the same shape. What figure am I?

   Ⓐ cube

   Ⓑ pyramid

   Ⓒ sphere

   Ⓓ cone

2. I am thinking of a number. When my number is multiplied by 2 it has a product between 10 and 20. Which is NOT my number?

   Ⓐ 9

   Ⓑ 8

   Ⓒ 6

   Ⓓ 4

Show What You Know

3. Monica had 69¢. She bought something at the store. Then she had 45¢ left. How much money did she spend?

   _____ ¢

   Explain how you found the answer.

   _____

   _____

   _____

   _____

4. Kip just loaded some games onto his computer. He almost doubled the number of games. Now he has 23 games. How many games did he have before?

   _____ games

   Explain how you found the answer.

   _____

   _____

   _____

   _____

336 three hundred thirty-six   **CCCXXXVI**   112 + 112 + 112   28 dozen

Chapter 15 Review/Assessment

NCTM Standards 1, 4, 5, 6, 7, 8, 9, 10

Which unit is best to measure the capacity of each container? Lessons 1 and 3

1.

   spoons   cups   pails

2.

   milliliters   liters

Match each amount to a container. Lesson 2

3. 2 cups

4. 4 cups

5. 16 cups

Which unit is best to weigh each object? Lesson 4

6.

   paper clips   books   bricks

7.

   paper clips   books   bricks

300 + 30 + 7   **CCCXXXVII**   three hundred thirty-seven   **337**

Is each real object *more* or *less* than 1 kilogram? Lesson 5

8.

_____ than 1 kilogram

9.

_____ than 1 kilogram

Match each object to its weight. Lesson 6

10.    2 tons

11.    2 ounces

12. 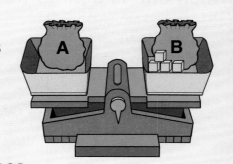   2 pounds

13. What temperature goes with the picture? Color the thermometer to show your estimate. Lesson 7

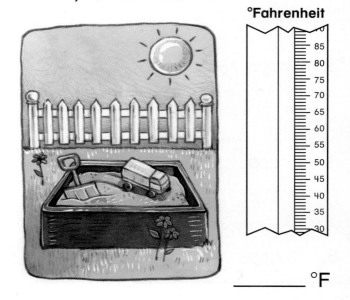

_____ °F

Problem Solving Lesson 8

14. Al put 20 cubes in two bags. He put more cubes in Bag A than in Bag B. He balanced the two bags by adding 4 cubes to the side with Bag B. How many cubes were in each bag?

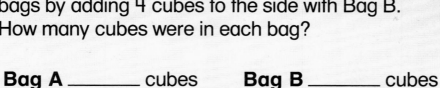

Bag A _____ cubes   **Bag B** _____ cubes

338 three hundred thirty-eight   CCCXXXVIII   169 + 169

Chapter 16: Multiplying and Dividing
Recording Different Ways

Name _____

You need
- connecting cubes

STEP 1 Making a Train

As a group, decide how many cubes to put in a train. Then each person makes a train of that size.
How many cubes are in your train? _____ cubes

STEP 2 Recording Trains and Cubes

How many trains did your group make? _____ trains
How can you find the number of cubes in all of the trains for your group?

_____

_____

_____

Write a number sentence to show what you did.

_____

STEP 3 Recording a Different Way

Can you think of other number sentences to describe the trains? Write as many sentences as you can.

_____

_____

_____

School-Home Connection

Dear Family,

Today we started Chapter 16 of *Think Math!* In this chapter, I will learn multiplication and division facts. There are NOTES on the Lesson Activity Book pages to explain what I am learning every day.

Here are some activities for us to do together at home. These activities will help me as I learn to multiply and divide.

Love,

_____

Family Fun

Fact Family Fandango

Work with your child to play this game. Your child will play a similar game in Lesson 3.

- You and your child each say a number from 0 to 10. Multiply the two numbers and write the sentence on a sheet of paper. For example:

    $6 \times 8 = 48$

- Have your child write all the number sentences that are part of the fact family that includes the sentence you just wrote.

    $8 \times 6 = 48,\ 48 \div 6 = 8,\ 48 \div 8 = 6$

- For every sentence your child records correctly, he or she gets a point. For every sentence he or she misses or records incorrectly, you get a point.

- Play several rounds. The player with more points at the end of the game wins.

Supermarket Math

Try this activity on your next trip to the store.

- Look for per-package counts on various products. For example the number of packets of oatmeal in a box, or the number of juice boxes in a pack. Look for packages that contain up to 10 items.

- Pose simple multiplication and division problems for your child to solve. If you find a 3-pack of juice boxes, then you might ask, "How many juice boxes are in 2 packages? What about 3 packages?"

- Invite your child to share how he or she found the answer.

Chapter 16 Lesson 1

Creating Multiplication Tables

NCTM Standards 1, 2, 6, 8, 9, 10

> Draw pictures of intersections, sets, or arrays to help you.

1. Fill in the table. Use your memory, draw a picture, or look back at work you have done before.

	0	1	2	3	4	5	6	7	8	9	10
×0	0		0		0						
×1											
×2											
×3			6		12		18				
×4											
×5	0										
×6											
×7											
×8											
×9											
×10											

NOTE: Your child is learning to complete a multiplication table using different strategies.

200 + 140 + 1   CCCXLI   three hundred forty-one   341

Part of a row from the multiplication table is shown.
What number is being multiplied in the row?

2.

| × 5 | 15 | 20 | 25 | 30 | 35 | 40 |

3.

| × ___ | 28 | 35 | 42 | 49 | 56 | 63 |

4.

| × ___ | 15 | 18 | 21 | 24 | 27 | 30 |

5. Jan knows that 2 × 9 = 18.
   How can she use this fact to solve 3 × 9?

   _____

   _____

6. How can you use 2 × 9 to solve 9 × 2?

   _____

   _____

Problem Solving

7. Why is the first row of the multiplication table on page 341 all zeros? Explain.

   _____

   _____

**Chapter 16
Lesson 2**

Multiplication and Division

NCTM Standards 1, 2, 6, 9, 10

4 + 4 + 4 = 12

Each picture is about the numbers 3, 4, and 12.

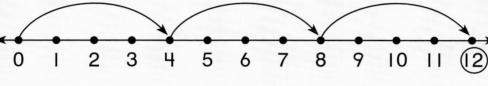

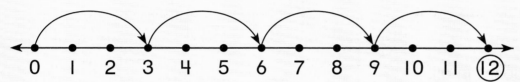

```
   3
   3
   3
+  3
-----
  12
```

What is missing? Complete each fact.

1.

$3 \times \underline{}4\underline{} = 12$

2.

$4 \times \underline{} = 12$

3.

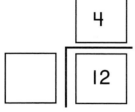

4.

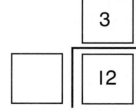

5.

```
     3
  × 4
  ___
  ▢
```

6.

```
     4
  × 3
  ___
  ▢
```

7.

$\underline{} \div 4 = 3$

8.

$12 \div 3 = \underline{}$

NOTE: Your child is learning to complete multiplication and division facts.

200 + 140 + 3   CCCXLIII   three hundred forty-three   **343**

**How many squares are there in all?
Write the missing numbers.**

9.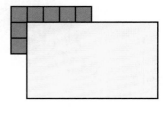

_____ × _____ = _____

10.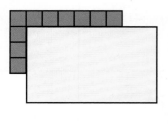

_____ × _____ = _____

11. There are 25 squares in all.
How many rows are there?

25 ÷ _____ = _____

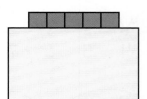

12. There are 24 squares in all.
How many columns are there?

24 ÷ _____ = _____

Challenge
Complete each fact family.

13. 4 × 5 = 20

   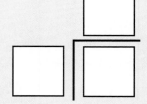   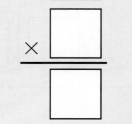

14. 3 × 6 = 18

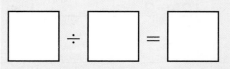   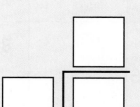   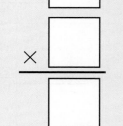

344   three hundred forty-four   CCCXLIV   172 + 172

Writing Multiplication and Division Fact Families

Chapter 16 Lesson 3

NCTM Standards 1, 2, 6, 8, 9, 10

What is missing? Complete each fact family.

1.

   $3 \times 6 = \boxed{}$ $\boxed{} \div 6 = \boxed{}$

   $\boxed{} \times \boxed{} = \boxed{}$ $\boxed{} \div 3 = \boxed{}$

2.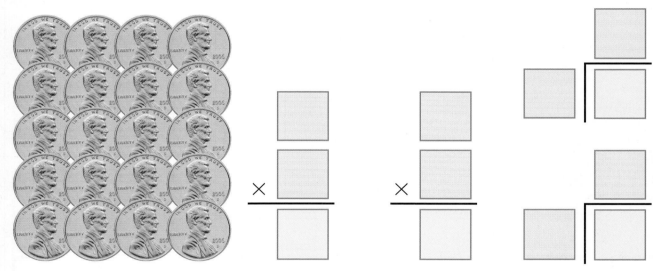

3.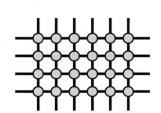

   $4 \times \boxed{} = \boxed{}$ $\boxed{} \div \boxed{} = \boxed{}$

   $6 \times \boxed{} = \boxed{}$ $\boxed{} \div \boxed{} = \boxed{}$

NOTE: Your child is learning to complete multiplication and division fact families.

What is missing? Complete each fact family.

4.

3 × ☐ = ☐

☐ × ☐ = ☐

12 ÷ ☐ = ☐

☐ ÷ ☐ = ☐

5.

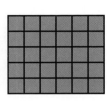

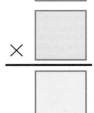

5
× ☐
───
☐

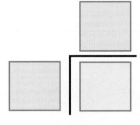

6.

3 × ☐ = ☐

☐ × ☐ = ☐

☐ ÷ ☐ = ☐

☐ ÷ ☐ = ☐

Problem Solving

7. Write a division story. Then write a number sentence to match it.

_____

_____

_____

346 three hundred forty-six   CCCXLVI   173 + 173

Chapter 16 Lesson 4
Connecting Pictures, Number Sentences, and Stories
NCTM Standards 1, 2, 6, 8, 9, 10

Draw a line from each picture to the matching number sentence. Complete the sentence.

1.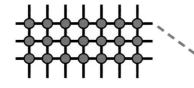

    $28 \div 7 = \underline{}$

2.

    $7 \times 3 = \underline{}$

3.

    $32 \div 4 = \underline{}$

4.

Sun	Mon	Tue	Wed	Thur	Fri	Sat
1	2	3	4	5	6	7
8	9	10	11	12	13	14
15	16	17	18	19	20	21
22	23	24	25	26	27	28

    $16 \div 4 = \underline{}$

5.

    $4 \times 10 = \underline{}$

NOTE: Your child is learning to write multiplication and division sentences for pictures.

200 + 140 + 7   CCCXLVII   three hundred forty-seven   347

Complete the fact families to match the pictures and stories.

6.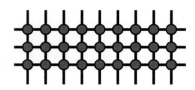

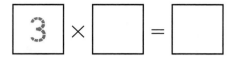

7. There are 24 marbles. They are shared equally among 3 children.

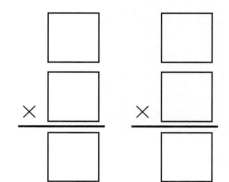

8. There are 36 baseballs. They are put into packages of 6 baseballs each.

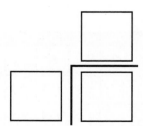

Challenge
Find each product.

9. $2 \times 3 \times 4 =$ _____

10. $6 \times 1 \times 5 =$ _____

Chapter 16 Lesson 5

Problem Solving Strategy
Act It Out

NCTM Standards 1, 2, 4, 6, 7, 8, 9, 10

Understand
Plan
Solve
Check

1. Three children equally share $1.50 in quarters. How much money does each child get?

   _____ ¢

2. A sheet of stamps has 6 rows with 5 stamps in each row. How many stamps are there in all?

   _____ stamps

3. 15 chairs are arranged in equal rows. How many chairs might be in each row?

   _____ chairs

4. Some children equally share 48 pretzels. Each child gets 8 pretzels. How many children share the pretzels?

   _____ children

 NOTE: Your child is exploring different ways to solve problems. Sometimes acting it out is an efficient way to solve a problem.

200 + 149   CCCXLIX   three hundred forty-nine

Problem Solving Test Prep

1. Cam has 4 coins in his pocket. He has only dimes and pennies. Which total amount is NOT possible for him to have?

    Ⓐ 13¢

    Ⓑ 21¢

    Ⓒ 22¢

    Ⓓ 31¢

2. There are 7 days in a week. Which number sentence shows how many days there are in 3 weeks?

    Ⓐ $7 - 3 = 4$

    Ⓑ $7 + 7 = 14$

    Ⓒ $7 \times 3 = 21$

    Ⓓ $4 \times 7 = 28$

Show What You Know

3. There are 49 erasers packed in boxes. The number of boxes is the same as the number of erasers in each box. How many erasers are in each box?

    _____ erasers

    Explain how you found the answer.

    _____

    _____

    _____

    _____

4. Each wagon has 4 wheels. How many wheels are there on 5 wagons?

Wagon	1				
Wheels	4				

    Explain how you found the answer.

    _____

    _____

    _____

    _____

Chapter 16 Review/Assessment

NCTM Standards 1, 2, 6, 8, 9, 10

1. Complete the multiplication table. Lesson 1

	0	1	2	3	4	5	6	7	8	9	10
× 3			6			15					
× 5					20						
× 7				21							

2. Sue knows that 8 × 5 = 40. How can she use this fact to solve 9 × 5? Lesson 1

_____

_____

What is missing? Complete each fact. Lesson 2

3.
$$7 \times \underline{} = 28$$

4.
$$\underline{} \div 7 = 4$$

5.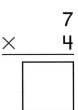

6.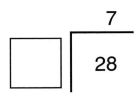

How many squares are there in all? Write the missing numbers. Lesson 2

7.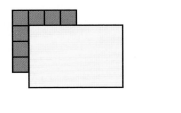

___ × ___ = ___

8.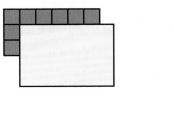

___ × ___ = ___

117 + 117 + 117   △   CCCLI   three hundred fifty-one **351**

9. What is missing? Complete the fact family. Lesson 3

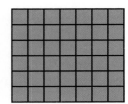

6 × ☐ = ☐   42 ÷ ☐ = ☐

☐ × ☐ = ☐   ☐ ÷ ☐ = ☐

Complete the fact families to match the stories. Lesson 4

10. There are 14 balloons. They are shared equally among 7 children.

☐ ÷ ☐ = ☐

☐ ÷ ☐ = ☐

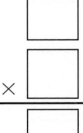

---

11. There are 81 books. They are put on shelves with 9 books on each shelf.

☐ × ☐ = ☐

Problem Solving Lesson 5

12. Five children equally share $1.00 in dimes. How much money does each child get?

_____ ¢

10 less
55, 45, 35, 25, 15, 5

Each number is **10 less** than the number before.

10 more
8, 18, 28, 38, 48, 58, 68

Each number is **10 more** than the number before.

add
To join 2 groups

addend

addends

addition
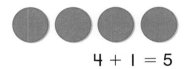

4 + 1 = 5

after
16 17 18 19 20 21 22 23 24 **25**

25 is **after** 24.

always

Green will **always** be picked.

area
The number of square units that cover a flat surface

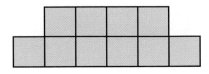

area = 10 square units

array
A rectangular arrangement of objects in rows and columns

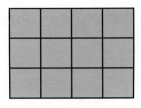

3 × 4 = 12

bar graph
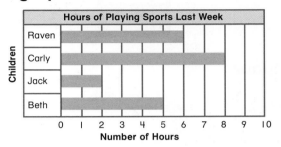

base
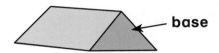

before
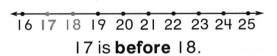

17 is **before** 18.

benchmark(s)
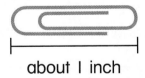

about 1 inch

between

30 is **between** 29 and 31.

biggest
115   15   5

115 is the **biggest** number.

Picture Glossary

353

bottom

calculator

capacity

centimeter

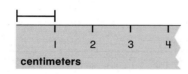

Used to measure the length of short objects

cent(s)

28 **cents**

certain

Green is a **certain** outcome.

change

I buy an apple for 50¢.   I pay with $1.00.   I get 50¢ **change**.

A line graph shows **change** over time.

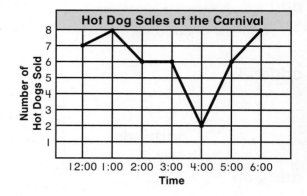

circle

circular

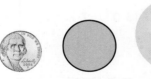

All of these have a **circular** shape.

close (to)

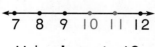

11 is **close** to 10.

closed figures

 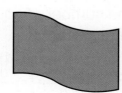

354

closer (to)

22 is **closer to** 20 than to 30.

coin(s)

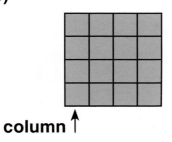

column(s)

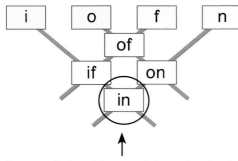

combination

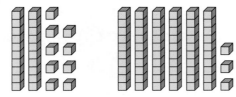

A **combination** of *i* and *n* is *in*.

compare

When you **compare** the two groups, you see that 29 is less than 63.

cone

congruent

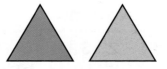

Figures that are the same size and shape are **congruent.**

corner(s)

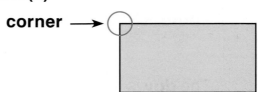

count

You **count** to find the number.

cube

cup

cylinder

data

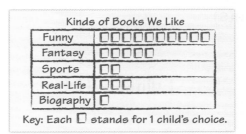

355

date

November						
Sunday	Monday	Tuesday	Wednesday	Thursday	Friday	Saturday
	1	2	3	4	5	6
7	8	9	10	11	12	13
14	15	16	17	18	19	20
21	22	23	24	25	26	27
28	29	30				

The **date** is November 28.

day

The **day** is Monday.

November						
Sunday	Monday	Tuesday	Wednesday	Thursday	Friday	Saturday
	1	2	3	4	5	6
7	8	9	10	11	12	13
14	15	16	17	18	19	20
21	22	23	24	25	26	27
28	29	30				

decimal point

$1.00
↑
decimal point

difference

6 − 4 = 2

The **difference** is 2.

digit

51
tens **digit** ↑↑ ones **digit**

51 has two **digits.**

dime(s)

1 **dime** = 10¢

division, divide

3 sets of 2

division fact

10 ÷ 5 = 2

$$4\overline{)12}^{\,3}$$

dollar(s)

$1.00

dollar sign

$1.00
↑ dollar sign

double

Both addends are the same in a **doubles** fact.

4 + 4 = 8

east

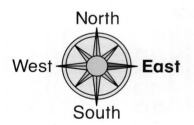

edge

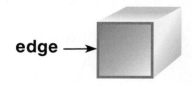

edge →

equal to

25 is **equal to** 25.

equally likely

Yellow and blue are **equally likely** to be pulled from the bag.

equivalent

The sets are **equivalent.**

estimate

To find about how many or how much

even

0, 2, 4, 6, 8, 10

exact

The **exact** time is 1:57.

face

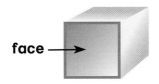

fact family

6, 7, and 13 are the numbers in this **fact family.**

6 + 7 = 13   7 + 6 = 13
13 − 6 = 7   13 − 7 = 6

factor

5 × 3 = 15
  ↖ ↑
  factors

Fahrenheit

fair share

feet

More than one foot

fewest

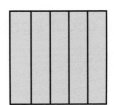

2 dimes and 4 pennies are the **fewest** coins that can be used to make 24¢.

fifths

357

flip

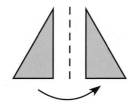

foot

A sheet of notebook paper is about 1 **foot** long.

1 **foot** = 12 inches

fourths

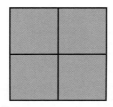

fraction

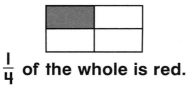

$\frac{1}{4}$ of the whole is red.

gallon

1 **gallon** = 4 quarts

gram

about 1 **gram**

graph

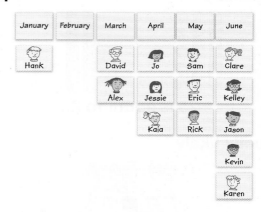

greater

Has more

greater than (>)

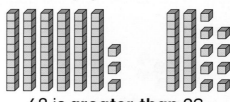

63 is **greater than** 29.
63 > 29

grid

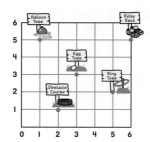

grow

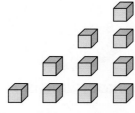

The pattern **grows.**

half

Half of the cubes are red.

half dollar

1 **half dollar** = 50¢

half hour

 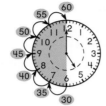

There are 30 minutes in a **half hour**.

halves

heavier, heaviest

The cherry is the **heaviest** object.
The cherry is **heavier** than the feather.

hexagon

horizontal

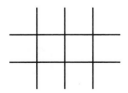

There are 2 **horizontal** lines.

hour(s)

There are 60 minutes in 1 **hour**.

hundreds

 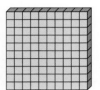

2 **hundreds** = 200

impossible

It is **impossible** to spin blue.

inch

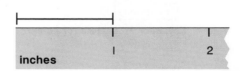

intersection

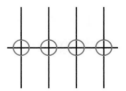

jump

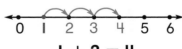

1 + 3 = 4

There are three **jumps** between 1 and 4.

359

key

kilogram

about 1 **kilogram**

labels

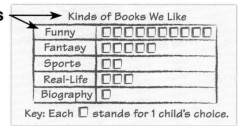

landing number

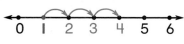

1 + 3 = 4

The **landing number** is 4.

landmark

The **landmark** time is 8:30.

least

The glass holds the **least**.

length

The spoon has a **length** of 6 inches.

less

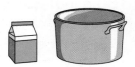

The milk box holds **less** than the pot.

less likely

You are **less likely** to pick yellow than red.

less than (<)

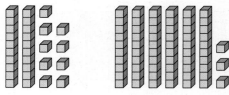

29 is **less than** 63.
29 < 63

level

The white cube is at **level** B.

lighter, lightest

A pencil is **lighter** than an apple. The pencil is the **lightest** object.

likely

Blue is a **likely** outcome.

line

line of symmetry

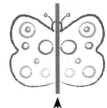

line of symmetry ↑

list

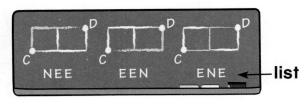

liter

1 **liter** = 1,000 milliliters

long

A pen is about 6 inches **long**.

measure

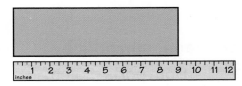

measurement

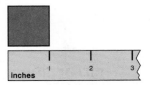

The **measurement** for the side is 1 inch.

mental math

Solving math problems without paper, a pencil, or a calculator

meter

Unit used to measure distances and the length of longer objects

middle

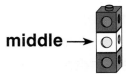

middle →

milliliter

1 **milliliter**

minute(s)

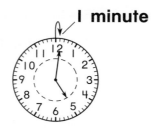

1 minute

mirror image, mirror line

Both parts match.

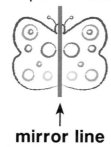

mirror line

missing factor

$9 \times ? = 18$

↑
missing factor

model

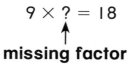

A **model** of multiplication

month

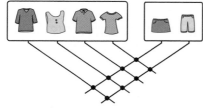

more

The pot holds **more** than the milk box.

more likely

You are **more likely** to pick green than blue.

more than

6 is **more than** 4.

most

The jar with the red lid has the **most** coins.

multiple

| Number of Eyes | 0 | 2 | 4 | 6 | 8 | 10 |

multiple choice

$$\begin{array}{r} 68 \\ +47 \\ \hline \end{array}$$

Ⓐ 115
Ⓑ 105
Ⓒ 100
Ⓓ 21

multiplication, multiply

5 sets of 3

multiplication fact

$5 \times 3 = 15$ $\quad \begin{array}{r} 2 \\ \times 6 \\ \hline 12 \end{array}$

multiplication table

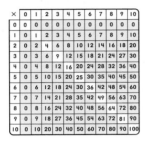

nearest

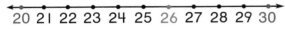

The **nearest** ten to 26 is 30.

never

Blue can **never** be picked.

nickel(s)

1 **nickel** = 5¢

noon

Noon
12:00 P.M.

north

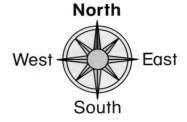

number line

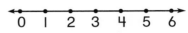

number pair

$2 + 8$
$5 + 5$
$7 + 3$

Each **number pair** has a sum of ten.

number sentence

$4 + 2 = 6$
$9 - 2 = 7$
$5 \times 2 = 10$

o'clock

The clock shows 10 **o'clock.**

odd

1, 3, 5, 7, 9

one dollar

$1.00

one fourth

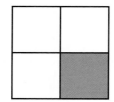

one half

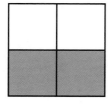

one third

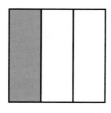

one twelfth

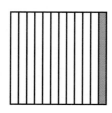

ones

2 **ones** = 2

open figures

order

0   30   58   61   85   100   302

These numbers are in **order** from the smallest to the biggest.

organized list

ounce

A slice of bread weighs about 1 **ounce**.

path

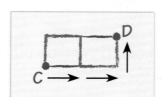

pattern(s)

30, 40, 50, 60, 70 . . .

pattern unit

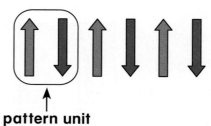

penny, pennies

1 **penny** = 1¢

pentagon

perimeter

The distance around a figure

2 centimeters · 2 centimeters · 2 centimeters · 2 centimeters

The **perimeter** measures 8 centimeters.

pictograph

picture graph

pint

1 **pint** = 2 cups

polygon

Each figure is a **polygon**.

possible

It is **possible** to pick a blue marble.

pound

A loaf of bread weighs about 1 **pound**.

predict, prediction

I **predict** the next block will be blue.

prism

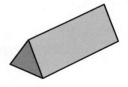

product

$4 \times 5 = 20$
↑
product

365

pyramid

quadrilateral

quart

1 **quart** = 4 cups

quarter

1 **quarter** = 25¢   **quarter** after 10

real-object graph

reasonable

99 × 6

600 would be a **reasonable** estimate.

rectangle

rectangular prism

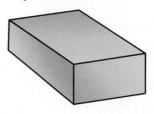

reflection

A **reflection** looks like the figure has been flipped.

regroup

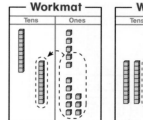 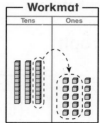

repackage

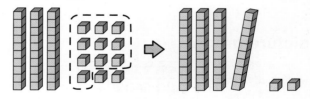

repeat

This pattern **repeats.**

rounding

Estimating to the nearest ten

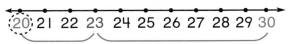

23 is closer to 20 than to 30.

row(s)

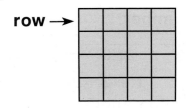

ruler

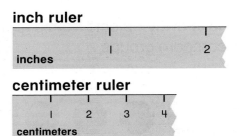

rule(s)

Add 2	
4	6
6	8
8	10

The **rule** is add 2.

same

 is the **same** as

second

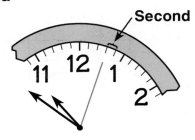

separate

You can **separate** the blue cubes from the red cubes.

set(s)

3 **sets** of 4

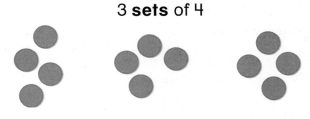

shortest path

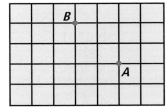

The **shortest path** from A to B is WWNN or NNWW.

side(s)

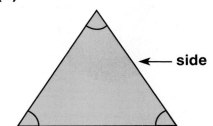

This figure has 3 **sides.**

367

similar

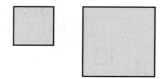

sixths

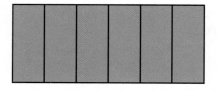

skip-count(ing)

460, 462, 464, 466, 468,...

smallest

115   15   5

5 is the **smallest** number.

solution

The way to solve a problem

south

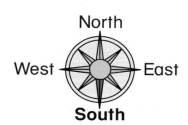

sphere

square

square unit

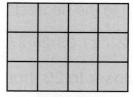

12 square units

starting number

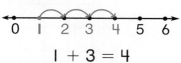

1 + 3 = 4

The **starting number** is 1.

subtract

To take away objects from a group or to compare groups

subtraction

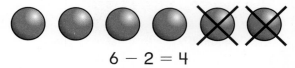

6 − 2 = 4

sum

6 + 3 = 9

The **sum** is 9.

symbol

☐ = 100 | or — = 10   • = 1

symmetry

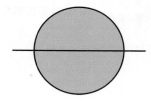

368

table

Our Favorite Places	
Place	Tally
arcade	||||| |||||
park	|||||
beach	||||| ||||| |||||

tall

This tower is four cubes **tall.**

tally

teens

The numbers 13, 14, 15, 16, 17, 18, and 19

temperature

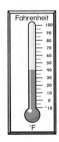

The **temperature** is 40°F.

ten(s)

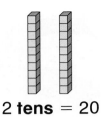

2 **tens** = 20

thermometer

thirds

time

The **time** is 3:00.

today

The day that is right now

tomorrow

The day that is after today

369

ton

A whale would be weighed in **tons**.

top

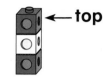

total

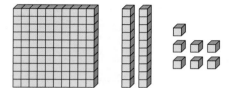

There is a **total** of 127.

total cost

total cost = $4.50

total value

total value = 50¢

towers

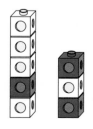

trade

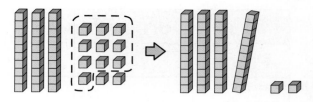

translate

The symbols **translate** to 124.

triangle

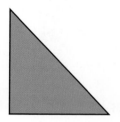

twelfths

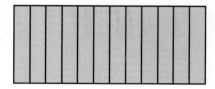

uncertain

Blue is an **uncertain** outcome.

units

The pen is about 4 **units** long.

unlikely

It is **unlikely** the spinner will land on yellow.

value

6**9**3

The **value** of the blue digit is 90.

vertex, vertices

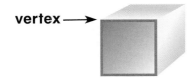

vertical

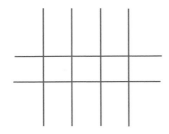

There are 4 **vertical** lines.

week

There are 7 days in 1 **week**.

weight

west

whole number

A number without a fraction
1, 2, 3, 4, 5, 6, 7, 8, 9 . . .

wide

The cube is about 1 inch **wide**.

word problem

If Jack jumped 5 times, and Jill jumped 10 times, how many times did they jump altogether?

yard(s)

1 **yard** = 3 feet

year

There are 12 months in 1 **year**.

yesterday

The day before today